CANADIAN MODERN

Seeing, Selling, and Situating Radio in Canada, 1922-1956

MICHAEL WINDOVER AND ANNE F. MACLENNAN

COMMISSIONING EDITOR: **MICHELANGELO SABATINO**
INTRODUCTION: **CHRISTINE MACY**

Dalhousie Architectural Press
Faculty of Architecture and Planning
Dalhousie University
Halifax, Nova Scotia, Canada
dal.ca/archpress

Editorial Board
Essy Baniassad, Chinese University of Hong Kong / University of Botswana
Sarah Bonnemaison, Dalhousie University
Brian Carter, SUNY Buffalo
Hans Ibelings, Architecture Observer
Christine Macy, Dean and Board Chair, Dalhousie University
Frank Palermo, Dalhousie University
Michelangelo Sabatino, Illinois Institute of Technology

Seeing, Selling, and Situating Radio, 1922-56
Series: *Canadian Modern*
ISBN 978-0-929112-70-1
Commissioning Editor: Michelangelo Sabatino
Publications Manager(s): Katie Arthur and Susanne Marshall
Designer: Anthony Taaffe
Cover photo: Peter Coffman
Printed by Halcraft Printers Inc.

© 2017 Dalhousie Architectural Press
All rights reserved. Published April 2017
Printed in Canada

Library and Archives Canada Cataloguing in Publication
MacLennan, Anne Frances, author
    Seeing, selling, and situating radio in Canada, 1922-56 / Anne F. MacLennan and Michael Windover.

(Canadian modern)
Includes bibliographical references.
ISBN 978-0-929112-70-1 (softcover)

    1. Portable radios--Canada--History--20th century. 2. Portable radios--Social aspects--Canada--History--20th century. 3. Advertising--Radios--Canada--History--20th century. I. Windover, Michael, author II. Title. III. Series: Canadian modern (Dalhousie Architectural Press)

K6548.C3M33 2017   621.3840971   C2016-907814-0

Contents

Exhibition information	5
Foreword: Michelangelo Sabatino	7
Introduction: Christine Macy	9
1. Seeing Radio: The Visual Culture of an Audio Medium	13
2. Selling Radio: How Radio Manufacturers Designed Desire	29
3. Situating Radio: How Radio Changed Canadian Space	65
Image credits	128
Acknowledgements	134
Notes on contributors	136

Front cover detail, Eaton's *Radio Catalogue*. 1935.

Exhibition information

This catalogue accompanies the exhibition "Seeing, Selling, and Situating Radio in Canada, 1922-1956" at Carleton University in the Discovery Centre located in the MacOdrum Library from 23 January to 30 April 2017, as well as "Making Radio Space in 1930s Canada" in the Carleton University Art Gallery from 27 February to 7 May 2017. It also accompanies exhibitions at the Allan Slaight Radio Institute at Ryerson University in Spring (May-June) 2017, at the Sound and Moving Image Library and Special Collections in the Scott Library at York University in Fall (September-December) 2017, and the Archives of Ontario from 1 September to 29 December 2017.

Front cover detail, Eaton's *Radio Catalogue*. 1936.

Foreword
Michelangelo Sabatino

Seeing, Selling, and Situating Radio in Canada, 1922-1956 is a co-authored study by an architectural historian and a communications historian who share a mutual interest in Canada's twentieth-century cultural, economic, social, and technological history. It is the fourth publication of our *Canadian Modern* series and, as with all of our previous publications, serves as a catalogue for an exhibition. The pedagogical power of exhibitions resides in their ability to narrate histories in textual, visual, and spatial terms. Unlike a book that reveals knowledge to the reader in a one-to-one relationship, exhibitions engage with the public in very different ways. Exhibitions, not unlike the medium of radio, are also an important tool for bringing scholarly research to a broader audience. While the textual, visual, and spatial components allow for the sharing of knowledge *in situ*, catalogues offer a more permanent research resource, especially for those who cannot personally visit the exhibition.

The title of this catalogue and accompanying exhibition do not refer explicitly to architecture or the natural and built environments of Canada. However, as the co-authors demonstrate, radio technology is inherently spatial—especially for a country such as Canada, characterized by its massive size, concentrated demography, and diverse geography. In 1936, the same year the Canadian Broadcasting Corporation was established, William Lyon Mackenzie King quipped in response to Benito Mussolini's invasion of Ethiopia: "…if some countries have too much history, we [Canada] have too much geography."[1] Similarly, Harold Innis, the innovative historian Marshall McLuhan credited as his foremost mentor, argued that Canada's national boundaries were "largely determined by the fur trade" and that "the present Dominion emerged not in spite of geography but because of it."[2] The vastness and diversity of Canada's geography, and its potentially alienating consequences, were overcome with a strategy of "technological nationalism" underlying the construction of the transcontinental railway – engineered by the Scottish-born Sandford Fleming. As Sir John A. MacDonald declared as the construction was about to begin in 1881:"The road will be constructed…and the fate of Canada, will then as a Dominion, be realized. Then will the fate of Canada, as one great body be fixed…"[3]

In terms of connecting Canadians from coast to coast, radio and the Trans-Canada Highway are to the twentieth century what the railroad and telegraph were to the nineteenth. It is worth recalling that in 1901, on the eastern most tip of Canada, St. John's Newfoundland served as the site of Guglielmo Marconi's first transatlantic wireless radio signal transmission. Just three decades later radio broadcasting was the nation's most powerful agent of spatial change. In their essays, Michael Windover and Anne F. MacLennan explore how radio became a household technology and the impact this early electronic medium had on Canadian perceptions and experiences of space. Their essays highlight the widening scope of the *Canadian Modern* series by including both a consideration of architectural projects—the architecture of broadcasting and transmission—as well as the space of reception: the interface of radio and advertising; radio and interior design; and radio and the practices of everyday life. The focus here is not on radio as a vehicle for architectural education for general audiences, which has seen some recent scholarly attention, but on the broad spatial and social effects of the radio itself.[4] With its comprehensive essays and compelling images, *Seeing, Selling, and Situating Radio in Canada, 1922-1956* will help stimulate further investigations on the transnational and global spatial implications of radio and on the creation of communities well before the world wide web dramatically reduced distances.

[1] See Dominion of Canada, *Official Report of Debates, Houses of Commons*, First Session: Eighteenth Parliament, Vol. IV, (June 18, 1936), 38-68. Quoted in Carl Guarneri and James Davis, eds., *Teaching American History in a Global Context* (London and New York: Routledge, 2008), 293.

[2] Harold A. Innis, *The Fur Trade in Canada* (University of Toronto Press, Torotno and Buffalo,1933) (1984 ed.), 393.

[3] Canada. House of Commons, *Debates*, 17 January 1881, p. 488. See Maurice Charland, "Technological Nationalism," *Canadian Journal of Political and Social Theory*, 10 (1986): 196-220.

[4] Shundana Yusaf, *Broadcasting Buildings – Architecture on the Wireless, 1927-1945* (Cambridge and London: MIT Press, 2014).

Working at CBC Radio

An interview with Sascha Hastings
Christine Macy

In January 2017, I had the opportunity to interview Sascha Hastings about her experiences at CBC. The interview format seemed especially apropos for an introduction to Michael Windover and Anne MacLennan's book *Seeing, Selling and Situating Radio in Canada*, allowing the reader to enter into the world of radio through the first-hand recollections of a radio producer. We talked about the distinctive language of radio production, the technology of ensuring high fidelity broadcasts over great distances, and the culture of CBC. The architect Peter Busby also joined us over lunch.

CM   In their book *Seeing, Selling and Situating Radio in Canada*, Michael Windover and Anne MacLennan look at 'radio space' in Canada – how radio was advertised, how people experienced it in the home, and how radio knit together the country at a very large scale — for example, with voices like Peter Gzowski that linked together Canada for 45 years across huge distances. And then of course, there's the architecture of the CBC building, like the one Philip Johnson designed in Toronto. So, what was it like to work at CBC, when you started out as an intern?

SH   When I started out as an intern, one of the big things that's different from now, is we still worked on reel-to-reel tape. They called it the 'Studer', that's the reel-to-reel machine and it's how you record. It's also how you edit. So I learned to edit with a razor blade. It's like the difference between drawing and computer rendering. So I'd edit with a razor blade and this little blue tape. And that meant when you're editing, you had to be really pretty precise. With computer editing, if you don't like it, you can hit 'undo'. But here, you cut, and pieces of tape end up on the floor. And then if something doesn't work you have to dig through the tape and find the piece on the cutting room floor. And I loved editing tape! It was so physical. You've got the two pieces of tape, and you reopen your headphones, and you listen, and you cut when somebody breathes in ... It's so sensual.

The other thing I liked with the reel-to-reel, is you had a real physical sense of where the beginning is and where is the end. And the interview is 'here'. It's in your hands. You can see it right in front of you. And you can guess how long the interview is at the end, because you can see how much tape is there (there's also a counter, so you actually know). That was wonderful, and I'm so glad that I was around just in time to learn that. We did that for a few years, and then we switched over to digital. Which is fine — I like it well enough and it gets the job done, and it makes it much easier to mix in music, or if you want to cross-fade anything, you can just click away. Whereas it used to be you had to do this in studio.

And the other thing is we had a lot more radio technicians back then. These guys (they were pretty much all guys) were real audio technicians. The producers were more about content, but these guys knew how to make it sound good. The producers, well — some are better than others at it. And so the techs would record all the interviews, they would voice track everything, they would mix the shows together. Voice tracking is when you lay down an introduction for example, or an extro, and then you add it afterwards. After you record the interview and edit it, the host would then maybe voice-track it afterwards, or you voice-track the whole show, which is all the bits between interviews.

CM   It almost sounds like tape itself has a spatial dimension.

SH   Exactly, and the other thing is now, you can do it all on your own. Then, you were working with a tech, working with another person. It was more social.

CM   Has that changed the way CBC is organized as a work space?

SH   Probably. I'm not in there very often. I still do some freelancing for them. I just aired an hour-long interview with John Neumeier, he's the artistic director of the Hamburg Ballet and choreographer. And I did all the work at home, I did all the research at home, but I actually wrote it in Italy.

CM   You're much more like an author then, working alone.

SH   Yes, I write the interview and Eleanor [Wachtel] actually does the interview (you know, records it), and I cut it when I get back. And we voice-track, and I package it, which means you put all the different components together. I could also get editing software and edit at home. But the editing software I know is the software that CBC uses. If I can get it all just by going into the CBC, I might as well.

CM   What were some of those other terms you talked about?

SH   Let's see ... there's a 'green'. A green is what we call a script. And that's because it used to be (I never saw these, they disappeared before I came along) an interview that was typed up, on a typewriter, on a triplicate mimeograph paper. The sheet that got ripped off —that's the one the host used — was green. They still call them greens. Or a 'three-way'. This is when you hook up three studios, say an interviewer in Toronto, a panelist in Halifax, and one in Vancouver. You could also have a 'four-way', a 'two-way', etc.

CM What's an 'end-to-ender'?

SH 'Double-ender'. The terms all sound slightly rude if you don't know the context! And we did this with John Neumeier. A double-ender goes like this. In our case, Eleanor was in Toronto, and he was in a studio in Hamburg. Actually, I won't do this with a studio example — we did that just as a backup. Sometimes you might do a 'phoner' with somebody — if you want studio quality, you record in studio in Toronto, and then wherever the guest is — let's say we did this at the Hamburg Ballet where they don't have a studio — she would have been on the phone with him, and while we would have recorded his interview at our end, the sound quality is terrible and you can't broadcast it. So we would have sent somebody in with a microphone, to record his voice onto a digital recorder in Hamburg. And then you send the audio to Toronto and then mix the two together. So you have the real tape from Toronto and the real tape from Hamburg, and you blend them. So she's hearing him over the phone and can respond and ask the next question, but in terms of having the broadcast quality sound — that's a double-ender. Sometimes we've also done that in studio where, for example, people in the Hamburg studio would record it there, because sometimes there are problems with the lines — just like with phone lines.

Sascha Hastings was born in Toronto, and studied German literature at the Universities of Toronto and in Freiburg, Germany. In 1998, she interned at the CBC for The Arts Tonight, and then became a producer, first on the book beat and then covering art and architecture. In 2005, at a time of CBC downsizing, she found herself out of a job and reinvented herself as the curator of Design at Riverside, Cambridge Galleries' space in Waterloo's School of Architecture. From there, she began her involvement with the Venice Biennale in Architecture, assisting with promotion of Philip Beesley's "Hylozoic Ground" in 2010; then managing for the RAIC, 5468796's "Migrating Landscapes" in 2012 and Lateral Office's "Arctic Adaptations: Nunavut at 15" in 2014. In 2016, she was co-principal of "The Evidence Room," which was in the central pavilion of the Architecture Biennale. Through this time, she continued to freelance for CBC Radio, producing original hour-long interviews with international arts figures in a special monthly feature called "Wachtel on the Arts" with host Eleanor Wachtel on "Ideas" (2007-16), and with Canadian writers for "The Next Chapter" with host Shelagh Rogers (2008-14). Currently, she is an independent arts producer, curator, editor, and architecture advocate in Toronto focusing on art and architecture.

Back cover detail, Eaton's *Radio Catalogue*. 1926.

Seeing Radio:
The Visual Culture of an Audio Medium

1

As radio became a prevalent household technology during the interwar years, it changed the acoustic environment of everyday life and became a visual focal point in homes and communities across Canada. The real and imagined "electronic hearth," which gathered Canadians together in small groups and larger, virtual publics, was usually a carefully chosen designed object, and one often purchased at substantial expense, especially in the 1920s and 1930s. While radio undoubtedly affected the soundscape of Canadian life, further embedding listening publics in electroacoustically experienced space, its entry into Canadian society was attended by a rich visual and material culture.[1] From the design of radio cabinets, which gave a representational quality to the new electronic medium, to advertisements aimed at selling equipment to enhance and afford radio listening lifestyles, radio was experienced not only in acoustic terms but visually and spatially.

In this book, we examine aspects of radio culture found in newspapers and magazines, in homes, and in the broader built environment, arguing that radio's visual and spatial dimensions were crucial to the formation of radio culture.[2] Our contention is that radio culture—a term we adopt to cover both things and practices associated with the electronic medium—was multi-sensorial and multifaceted. As a social instrument, radio facilitated the creation of a new kind of public space. And the imaginative potential in visual culture and design helped shape the contours of this social space, which became deeply embedded within Canadian society by the time television altered the visual-acoustic landscape again in the 1950s.

Household technologies do not simply emerge. Their successful adoption and their impact on culture is contingent on a series of different actors and factors, including manufacturers and users, older technologies and patterns of use, advertisers and merchandisers, and networks of distribution and consumer habits. This book— which draws material mainly from the popular press across the country, as well as the T. Eaton Company Fonds housed at the Archives of Ontario, the Canadian Broadcasting Corporation Fonds at Library and Archives Canada, and the trade literature and artefact collections of the Canada Science and Technology Museum—illustrates many of these agents. Radio provides an interesting case; the size, shape, design, usual location, and use of radio sets evolved from the 1920s to the 1950s as radio firmly inserted itself into the daily lives of Canadians. Radio's role in the home was set apart from the many household appliances that were purchased widely as part of postwar consumption; its size, design, price, and function made it an early and distinct addition to the home. The generation and successful incorporation of radio culture into everyday life depended on visual representation and material infrastructure. An aim of this book as a whole is to demonstrate how important visual and material culture is to understanding how radio became a crucial part and affected the formation of modern Canadian society.

Imagining Radio at Home
Radio began as a hobbyist's pursuit, with a network of technologically engaged radio enthusiasts establishing the first audience.[3] Indeed, many men received training as radio operators during the First World War and, upon their return home, continued the hobby and formed the backbone of radio station operations with their skills. Makeshift sets were commonly assembled by boys and young men, who were enthralled with the possibility of travelling great distances virtually, encountering strange voices that were pulled in from far flung points around the globe.[4] Radio moved very quickly out of attics, garages, and the barns of the hobbyists in countries such as England, where radio reception range covered most of the country at the inception of the British Broadcasting Company in 1922, and the United States, where a large population and strong commercial incentive established three large networks by 1928.[5] While not as strong as that in the United States and England, the radio market in Canada burgeoned quickly. In the 1920s, radio broadcasting ceased to be a novelty and the new medium started to enter the homes of most Canadians. By 1938, 84 percent of Canada was in radio signal range, the level reached by England in 1922.[6] Sparse radio coverage of large areas, lack of electrification, and the price of the early radios made the radio receiver an aspirational object for most Canadians, who continued to use their homemade crystal sets well into the 1920s. By

Fig 1.1

the mid-1920s, the radio, encased in fine furniture, became the electronic apparatus found at the heart of well-to-do Canadian households.

Advertisements for radio sets began appearing in Canadian cultural magazines and newspapers before radio was a commonplace fixture in Canadian domestic interiors. In 1922, for instance, an advertisement for a Marconi Long Range Receiver depicted a well-heeled group of men and women listening and looking at a radio set complete with speaker (Fig. 1.1). Above them is pictured a world of entertainment and information now available over the airwaves. "Radio," claims the copy, "furnishes the home with constant contact with the world and its doings." Here, the unique spatial experience of radio is highlighted, but so too, if more subtly, is the idea of the radio apparatus as home furniture. The slightly later Marconiphone II (1923-24) from the Canada Science and Technology Museum collection gives a sense of the kind of instrument being wondered over in the advertisement (Plate 1.1). The crisp mahogany and Bakelite box containing the radio receiver has the appearance of a scientific instrument, while the metal and walnut speaker horn resembled the gramophone familiar to contemporary listeners. Though the advertisement suggests "constant contact", the reality was not so. In Canada, the first commercial radio licence was only issued in 1922 for the Marconi Company's own station in Montreal, which commenced as experimental station XWA in 1919, and while there were some established stations in major cities at this time, round the clock programming was not the norm and would not be for more than a decade later.[7]

Fig 1.1 Advertisement for Marconi. *Montreal Gazette*. 1922.

Representing Radio Space

The early Marconi advertisement underlines how important advertising was in creating a market for this new technology. Indeed, advertisers, retailers, and manufacturers had an enormous impact on the creation of radio culture. An early supporter was Canada's largest retailer, the T. Eaton Company. Once a dry goods store based in Toronto, by the mid-1920s Eaton's was a major operation with flagship department stores in Montreal, Toronto, Winnipeg, and Calgary and a vast distribution network that penetrated deep into the Canadian hinterland, fuelled by its successful mail order catalogues. Ever conscious of trends in retailing, the company established a radio department in 1921 and even applied for a broadcasting licence to demonstrate its products in Toronto.[8] A similar service was set up in the Winnipeg store in April 1922.[9] While the company only operated the radio stations until 1926, the presence of wireless broadcasting in the department store evinced not only the company's engagement with new technologies in a retail environment (like Wanamaker's south of the border)[10] but the important links between selling radio equipment and the idea of radio as a medium. Sold as part of a tasteful, modern living room, companies such as Eaton's helped foster perceptions of radio for its consumers. Unsurprisingly, it was just at the moment when radio receivers became a more common sight in Canadian living rooms that Eaton's left the broadcast business and focussed on selling (and manufacturing) receivers.

A radio catalogue produced for the company's western mail order business based in Winnipeg in the year it left broadcasting (1926-27) indicates how Eaton's sought to represent the new medium (Plate 1.2).[11] The front cover depicts a young family around a rather ornate lowboy console complete with a speaker placed on top. A fashionable mother is seated in a comfortable armchair, suggestive of a living room, while her son sitting on her lap and her husband standing nearby look with happy astonishment at the radio set. Behind them, French doors open onto a balcony and night sky, which is filled with a spectral orchestra. The message is clear: through radio, you can listen to the symphony at home. This window to the world of high culture — the radio set —is available for purchase through mail order.

The back cover adds to the story, helping new Canadian consumers imagine the new technology and offering a sense of the range of spaces created by it (Plate 1.3). The circular speaker seen on the front cover is reproduced here and around its circumference are pictured activities that could be participated in virtually through the radio, much like the orchestra introduced on the front. Listeners, like the family pictured on the bottom left, could tune into a variety of different spaces, which would, in a way, transform their own living room into the bleachers at a ball game, seats in a concert hall, the floor of a ballroom or the interior of a church. The arrangement of activities around the speaker alludes to the face of a clock, perhaps suggesting activities that take place at different times. The regimentation and compression of time and space, which is deeply engrained in the experience of modernity, underlines here the complexity of radio space.

Space is further represented on the back cover in the depiction of wireless broadcasting. Flanking the circular speaker, transmission towers disseminate radio signals, enabling the miracle of broadcasting. This detail reminds us of the contemporary fascination with the infrastructure of radio, a topic taken up in the final chapter of this catalogue. Finally, on the lower right we see a broadcaster speaking into a microphone, in a space furnished like a living room. The continuity of space represents radio as an intimate and non-threatening technology. Obviously, Eaton's wanted to increase the number of radio users so it could sell more equipment.

The intimacy of radio is significant. Radio historians such as Michele Hilmes have highlighted that intimacy was recognized as a key feature of the medium.[12] For example, she and Jason Loviglio note that Anning S. Prall, the first chief commissioner of the Federal Communication Commission in the United States, explained in 1936 that radio "claims a more intimate relationship with the public."[13] Radio could construct publics that were simultaneously intimate and national.[14] It had a neighbourly quality and the catalogue cover points to this sense of familiarity in the radio address. The image might also suggest agency on the part of listeners, who, like an auditorium audience, participate in activities by listening.[15] In this rendering, the broadcaster is in the audience's space and yet still outside. The covers of the *Eaton's Radio Catalogue* succinctly express the spatial complexity of radio culture at an early moment in the domestication of the technology.

Looking at Radio

An advertisement for the Canadian General Electric Radiola 32 from the March 1928 issue of *Canadian Homes and Gardens* tells us something about the development and domestication of radio technology and provides clues about the cultural connotations of the new medium (Fig. 1.2). This was an expensive model ($1250 in 1928) and was sold as a "fine" instrument. The average male worker's annual salary in 1931 was $927, placing these radios out of reach for the average Canadian.[16] Salary groups in 1931 ranged from public administration professionals with the highest average annual salary of $2,990, while agricultural workers earned the lowest annual average salary of $319.[17] The image in the advertisement reveals little about the home, showing only the corner of a room and few hints about its occupants. The availability of radio stations concentrated in Canadian cities, combined with the fact that urban centres far outpaced rural areas

Fig 1.2

Fig 1.2 Advertisement for Radiola. *Canadian Homes and Gardens.* 1928.

in wealth generation, suggests this radio's home was mostly likely in a city. The copy for the advertisement describes the "custom-built" console, its technological advantages (alternating current-powered eight-tube Super Heterodyne circuit, rotatable antenna loop, and built-in speaker), and, importantly, points to its multi-sensorial qualities: "Delight your finer senses by *seeing* Radiola 32 and *hearing* it in your home".[18] Much like the newly restyled Ford Model A, appearance was recognized by designers and marketers as a critical part of situating technologies in modern lifestyles.[19]

The Radiola 32 is staged in a den or living room, where contrasting themes of home and travel are made apparent through the arrangement of decorative features in the room. The radio console is opened to illustrate the knobs and dials of this refined, domestic technology, and the power cord, which follows the angle of the stairs to an outlet, highlights the modernity of the instrument. With A/C power, this radio does not require the often messy batteries necessary to power many models in use in Canada. This receiver is thus designed specifically for the public areas of the home. Its somewhat awkward placement at the foot of stairs sets up a counterpoint to the rustic brick fireplace (which provides an indication of its size). The idea of radio as the electronic, modern hearth is evident through this contrast. The fireplace, fuelled by wood, is topped with a mantel containing plants and a model of a tall ship. The radio, powered by electricity, holds a vase with pussy willows, whose forms might be read as evoking radio waves entering the console below. On the wall above is a painting of a steamship. The allusions to travelling the waves (now airwaves) links the radio to the excitement of discovery, while the warmth and stability of the furniture associates the receiver firmly with traditional Canadian values of home. This advertisement for the Radiola 32 points to the negotiation of new possibilities heralded by modern technologies with the well-entrenched notions of middle-class (more likely upper-class) family life and home.

By staging the Radiola 32 in this space—representatively powerful yet entirely impractical—the advertisers reinforce connotations associated with radio. After all, as the copy asserts, the Radiola 32 is a "conception of what radio should be". A strong sense of the cultural capital of the radio as a possession is further indicated in the text of the advertisement as it extolls the virtues of the Radiola 32 as "this custom-built receiver…[with] refinements of appearance and performance which inspire true pride of ownership."[20] The goal here is the domestication of the technology and the representation of the multivalent modernity of radio culture. The radio is framed as a prestigious addition to modern lifestyle, and with its explicit comparison to the hearth, the model is imagined as an extension of the kind of space located at the centre of wealthiest Canadian homes and society.

While radio was first commercially available in Canada in 1922, 1928, when this advertisement for the Radiola 32 appeared, was an important year of change for the medium in North America. After that, sophisticated radios with tuning dials were increasingly needed in Canada. The Royal Commission on Radio Broadcasting, more commonly known as the Aird Commission, commenced its investigation in 1928, and, perhaps not coincidentally, that same year also marked the North American Radio Broadcasting Agreement. This agreement both solidified CBS and NBC's Blue and Red Networks and reassigned radio frequencies across the continent to reduce interference with listeners' enjoyment of their favourite radio stations in Canada, the United States and Mexico.[21] All was not rectified in the late 1920s; however, by 1937, with the Havana American Regional Broadcasting Agreement, all clear channels were to be occupied by fifty-kilowatt stations. The erection of the CBC's first high-power transmission stations in Verchères, QC and Hornby, ON, coincided with this accord, acting as a kind of assertion of sovereignty.[22]

The reassignment of radio frequencies changed radio for the listener and made the purchased radio receiver crucial to listening. As radio stations multiplied, the ability to tune into more than one station in the same region became essential. This reallocation of radio frequencies overrode the previous reallocation of February 20, 1925, which was the result of The Gentleman's Agreement of 1924 with the United States. That agreement was regulated in Canada by the Radio Branch of the Department of Marine and Fisheries. The practice until 1928 effectively banned simultaneous broadcasting by assigning only one wavelength to be shared in each Canadian city, with the exceptions of Montreal, Toronto, and Vancouver.[23] The reassignment of the wavelengths made technological improvements in the design of radio receivers important to the listeners. The changes begun in 1928 started the move away from one shared wavelength per Canadian city in order to keep up with an increasingly competitive North American broadcasting environment spurred by changes in the United States.[24] The establishment of the Federal Radio Commission in 1926, the passage of the American Radio Act in 1927, and subsequent rearrangement of radio frequencies in North America made the purchase of a radio receiver in Canada more important in cities and areas with competing radio stations; a crystal set was no longer sufficient.

By 1936, the market was flooded with radios of different sizes, capabilities and design, and the role of radio and its place in the home shifted as listening patterns changed and new models became available for purchase. Westinghouse's Blue Ribbon Air Pilot Radio was advertised for its ability to "split hairs" (Fig. 1.3). It was "twice as selective as previous receivers ... wherever you live there are probably two powerful

Fig 1.3

stations so close together that even the best of radios in the past have been unable to separate them."[25] Again, spatial proximity is a key feature of radio development. With more powerful stations came clearer reception, but also more difficulty in tuning. For instance, listeners wanting to tune in to Toronto might find themselves in Cincinnati instead. The Blue Ribbon Air Pilot was a floor model, despite the growing popularity of less expensive and very popular tabletops. This floor model was loaded with extra features and capabilities that made it worth the added expense and earned it a place in the living rooms of Canada's finer homes. The console was a status symbol, something to be admired not only for its technological capabilities but its appearance and, as discussed in the third chapter, floor models like this were indicators of modernism's (or Art Deco's) acceptance in mainstream interior design schemes.

The Synesthetic Experience

Radio had a life of its own in print media. While newspapers felt threatened by radio in its early stages immediately after the First World War, this eased when it was discovered that radio stations were expensive to run. Instead of refusing to list radio programs in their pages, newspapers welcomed program listings and introduced regular columns

Fig 1.4

commenting on the new technology, as well as publishing radio advertising in ways that highlighted the potential of the medium and that translated the aural into the visible. A 1930 advertising campaign for Philco radios illustrates how acoustic clarity could be depicted visually (Fig. 1.4). The strategy is simple yet effective: two photographs of Maurice Chevalier—one blurry, the other sharp—are presented side by side. Above the blurry one, the text reads: "*Un*balanced radio means distorted tone." Above the sharp one: "Balanced radio means true clear tone." Given the technological progress made by the Philco Company, consumers might expect from its products a better, more naturalistic listening experience. In this advertisement, the consumer is being asked to visualize what they might hear. The visual portrayal of the radio provided incentive for the transition from crystal set to the floor model and eventually the battery-operated, less expensive table top by 1931.

The inclusion of a radio personality, in this case a star of "The Love Parade," indicates another way print media added a dimension to radio culture. The "voice full of personality" is given a face, just as the radio is given a face with a Philco console. In both cases, radio content is framed and represented. The image of Chevalier for instance, gives visual form to what might have been an anonymous voice, influencing the way listeners imagined what they heard. Similarly, the design of the radio cabinet gave radio reception a tangible form. Like the Radiola 32 or the later Westinghouse Blue Ribbon, the wooden case conferred an air of respectability and familiarity on the medium. For listeners gathering around a Philco console, radio looked like this.

Fig 1.3 Advertisement for Westinghouse. *Maclean's Magazine*. 1936.

Fig 1.4 Advertisement for Philco. *Maclean's Magazine*. 1930

Fig 1.5

An advertisement for the Northern Electric Mirrophonic Radio series takes the visual analogy even further (Fig. 1.5). In this instance, purity of sound is visually represented with a mirror. The copy points to the larger, modern electroacoustic soundscape, noting that the "Mirrophonic tone chamber" of the cabinet was "designed by the men who developed Mirrophonic Sound Systems for Moving Picture Theatres." At the movies, film and audio must necessarily mirror each other; at home, precision and clarity of sound help the listener to form a mental image of the performance. The Mirrophonic Radios exemplify not only the advertiser's art in visually representing the development of an acoustic medium, but also the company's desire to reinforce these visual analogies in naming, designing, and marketing the consoles. The Mirrophonic was another way of making the aural visual.

While Philco's and Northern Electric's advertising campaigns used visual analogies to emphasize precision and accuracy, other advertising explored the colour of sound. The 1944 pamphlet "Your Coming Radio as forecast by General Electric" examines the emergent technology of frequency modulation (FM) (Plate 1.4). The CBC opened experimental FM transmitters in Montreal (one for English, one for French) and Toronto in 1946,[26] and several private FM stations opened in the years following the war, such that gradually FM became established as another radio medium. General Electric's pamphlet extolls the virtues of the FM radio that will "bring you natural color music…a Fidelity Miracle." New radio equipment, now paired with record players in low-rise rectangular cabinets, once again returned to a place of honour in the home, with a focus on high definition listening practices in the home entertainment spaces of Canadians.

Colour was also used in the design of radio cabinets made of Bakelite, Catalin, and other early plastics in the postwar era. A 1951 Crosley advertisement for the "Coloradio" is a good example (Fig. 1.6). Suggesting that the small radio "brings out the artist in you", the female owner is cast as an interior designer, using the "color-rich cabinet" as part of her palette for designing a tasteful home. With models in "ebony, chartreuse, maroon, aqua, green and white…there's a color to harmonize with the decoration scheme of every room", the copy asserts. The radio in the postwar era, having played such a significant role in bringing home news during the war, has now become a part of everyday life, resonating with the visual and material culture of the home as effortlessly as a listener can tune into her favourite station.

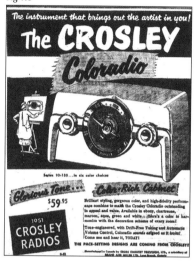

Fig 1.6

Conclusion

Advertisements and the design of radio receivers played a key role in defining radio. Indeed, the visual and material culture of radio was essential to generating, elaborating, and even speculating on the meaning of the medium. Looking closely at advertisements in daily newspapers and popular magazines; catalogues and promotional material by manufacturers; the design of radio consoles; and the architectural framing of broadcasting, transmission, and reception (at home or on the road) underlines the imbrication of factors that affected the formation and development of radio culture.

We pursue this further in the next chapter, which explores more extensively the marketing of radio. Advertisers introduced future listeners to radio through newspapers and magazines, where readers could see radio sets that they might never see in their own homes or towns. These radios created an imagined space in which Canadians could aspire to acquire a radio of their own and, eventually, a newer model, as radio functions and forms multiplied.

In the final chapter, we return to radio space. Here, we explore radio in a series of interpenetrating networks of commerce, politics, and technological development. The radio creates space at home and helps introduce modern design into the living room, becoming an agent of spatial change and interior design. The architecture of radio broadcasting facilities contributes to the formation of national, regional, and local radio space. Ultimately, radio becomes an indispensible feature of the post-war home, and a highly mobile technology. As seen in the 1922 Marconi advertisement discussed at the outset of this chapter, radios offered a connection to the wider world from seemingly anywhere in Canada.

Fig 1.5 Advertisement for Northern Electric. *Montreal Gazette*. 1937.

Fig 1.6 Advertisement for Crosley. *Winnipeg Free Press*. 1951.

Plate 1.1 Marconiphone II, c. 1923-24.

Plate 1.2 Front cover of Eaton's *Radio Catalogue*. 1926.

Plate 1.3 Back cover of Eaton's *Radio Catalogue*. 1926.

Plate 1.4 General Electric pamphlet. 1944.

Notes

[1] See Emily Thompson, *The Soundscape of Modernity: Architectural Acoustics and the Culture of Listening in America, 1900- 1933* (Cambridge, MA: MIT Press, 2002). The term "soundscape" is derived from R. Murray Schafer's work and has been influential in sound studies. R. Murray Schafer, *The Soundscape: Our Sonic Environment and the Tuning of the World* (Rochester, Vermont: Destiny Books, 1977). For more on this field, see Jonathan Sterne, ed., *The Sound Studies Reader* (London: Routledge, 2012).

[2] The major exception is Bill McNeil and Morris Wolfe, *Signing On: The Birth of Radio in Canada* (Toronto and New York: Doubleday, 1982). The term 'Radio Cultures' has been used before: Michael C. Keith, ed., *Radio Cultures: The Sound Medium in American Life*. (New York: Peter Lang, 2008) and used to make reference to the inclusivity in the sense that radio offers a space for many voices, communities and ideas. Similarly, Lizabeth Cohen brings to light the contribution radio makes to the home while not disturbing its existing ethnic culture. Lizabeth Cohen, *Making a New Deal: Industrial Workers in Chicago 1919-1939* (Cambridge: Cambridge University Press, 1990). Our sense of radio culture refers to the commonalities of radio in the home and the use of space centred around it as novel in the 1920s, but enduring as it was transformed with each decade.

[3] Anne F. MacLennan, "Learning to Listen: Becoming a Canadian Radio Audience in the 1930s," *Journal of Radio & Audio Media*. 20 (2013): 324.

[4] Extensive research by Michele Hilmes indicates that the "ability to escape the determinations of gender" made it very popular among their female counterparts Michele Hilmes, *Radio Voices: American Broadcasting, 1922-1952* (Minneapolis: University of Minnesota Press, 1997), 132-136; Anne F. MacLennan, "Circumstances Beyond Our Control: Canadian Radio Program Schedule Evolution during the 1930s" (PhD diss., Concordia University, 2001): 2-3.

[5] Robert W. McChesney, *Telecommunications, Mass Media, and Democracy: The Battle for the Control of U.S. Broadcasting, 1928-1935* (New York: Oxford University Press, 1993); Christopher H. Sterling and John M. Kittross, *Stay Tuned: A Concise History of American Broadcasting* Second Edition (Belmont, California: Wadsworth Publishing Company, 1990), 66-68, 108-110.

[6] Canadian Broadcasting Corporation, *Canadian Broadcasting: An Account of Stewardship*, excerpted from Canada. Parliament. House of Commons. Special Committee on Radio Broadcasting. *Minutes of Proceedings and Evidence, Nos. 2 and 3* (Ottawa: King's Printer, 1939), 11 as cited in Anne F. MacLennan, "Circumstances Beyond Our Control: Canadian Radio Program Schedule Evolution during the 1930s" (PhD diss., Concordia University, 2001): 30-31.

[7] Pierre C Pagé, "L'origine des stations XWA (1915) et CFCF (1922) de Marconi Wireless Telegraph: des données historiographiques à corriger," *Fréquence/Frequency* 5-6 (1996): 151-168; Donald G. Godfrey, "Canadian Marconi: CFCF the Forgotten First," *Canadian Journal of Communication* 8, (September 1982): 56-71; Gilles Proulx, *La Radio d'hier et d'aujourdhui* (Montreal: Editions Libre Expression, 1986), 27; Gilles Proulx, *L'Aventure de la radio au Québec* (Montreal: Editions La Presse, 1979), 22.

[8] Mary Vipond, *Listening In: The First Decade of Canadian Broadcasting, 1922-32* (Montreal: McGill-Queen's University Press, 1992), 18. The station CJCD opened in 1922 with a 15 watt transmitter. It was increased to 100 watts the following year. It was closed down in 1926. See Canadian Communication Foundation website, Station History, http://www.broadcasting-history.ca/index3.html?url=http%3A//www.broadcasting-history.ca/listings_and_histories/radio/histories.php%3Fid%3D811%26historyID%3D508.

[9] Under the direction of Lynn V. Salton, Eaton's ran CKZC Winnipeg with 420 metres and 100 watts for a short period of time. Salton also headed up the Radio Catalogue division, according to the Canadian Communication Foundation website.

[10] Noah Arceneaux, "Wanamaker's Department Store and the Origins of Electronic Media, 1910-1922," *Technology and Culture* 51 (Oct 2010): 815.

[11] See Michael Windover, "Placing Radio in Sackville, New Brunswick," *Buildings & Landscapes* 24, (Spring 2017) (forthcoming).

[12] Hugh Chignell, *Key Concepts in Radio Studies*. (London: Sage, 2009); Andrew Crisell, *Understanding Radio, Second Edition* (London: Routledge, 1994); David Hendy, *Radio in the Global Age* (Cambridge: Polity Press, 2000); Hilmes, Michele. *Radio Voices: American Broadcasting, 1922-1952* (Minneapolis: University of Minnesota Press, 1997); Susan Smulyan. *Selling Radio: The Commercialization of American Broadcasting, 1920-1934* (Washington: Smithsonian Institution Press, 1994); Jason Loviglio, *Radio's Intimate Public: Network Broadcasting and Mass-mediated Democracy*. Minneapolis: University of Minnesota Press, 2005; Len Kuffert, "'What do you expect of this friend?': Canadian radio and the intimacy of broadcasting," *Media History* 15, (August 2009): 303-319.

[13] Mutual Broadcasting System's inaugural coast-to-coast broadcast, September 1936, University of Memphis Radio Archive as cited in "Introduction" in Michele Hilmes and Jason Loviglio, eds., *Radio Reader: Essays in the Cultural History of Radio* (New York: Routledge, 2001), xi.

[14] See Michael Windover, "Designing Public Radio in Canada," *RACAR* 40, no. 2 (2015): 42-56.

[15] Kate Lacey, *Listening Publics: The Politics and Experience of Listening in the Media Age*. (Cambridge: Polity Press, 2013); Anne F. MacLennan, "Learning to Listen: Becoming a Canadian Radio Audience in the 1930s," *Journal of Radio & Audio Media*. 20 (2013): 311-326; Anne F. MacLennan, "Reading Radio: The Intersection between Radio and Newspaper for the Canadian Radio Listener in the 1930s," in *Radio and Society: New Thinking for an Old Medium*, ed. Matt Mollgaard, 16-29 (Newcastle upon Tyne: Cambridge Scholars Publishing, 2012).

[16] Canada. Dominion Bureau of Statistics. *Seventh Census of Canada*, 1931 Volume 5 (Ottawa: King's Printer, 1950), Table 14, 28.

[17] Canada. Dominion Bureau of Statistics. *Seventh Census of Canada*, 1931 Volume 5 (Ottawa: King's Printer, 1950), Table 14, 28.

[18] Radiola Canadian General Electric Co., Limited, "The Ultimate in Fine Radio," *Canadian Homes and Gardens* (March 1928): 4.

[19] For a perspective on the role of appearance in modern design and lifestyle, see Michael Windover, *Art Deco: A Mode of Mobility* (Québec: Presses de l'Université du Québec, 2012).

[20] Pierre Bourdieu, *Distinction : A Social Critique of the Judgement of Taste*, trans.Richard Nice (Cambridge, Mass.: Harvard University Press, 1984); Radiola Canadian General Electric Co., Limited, "The Ultimate in Fine Radio," *Canadian Homes and Gardens* (March 1928): 4.

[21] Anne F. MacLennan, "Crossing the Border: The Case of CBS, NBC and Mutual affiliate stations outside the United States," Paper presented at Saving American's Radio Heritage Conference, Radio History Task Force Conference, 26 February 2016, Washington, D.C.; Anne F. MacLennan, "Editor's Remarks: Transcending Borders" Reaffirming Radio's Cultural Value in Canada and Beyond," *Journal of Radio and Audio Media* 23 (2016): 197-199.

[22] Michael Windover, "Transmitting Nation: 'Bordering' and the Architecture of the CBC in the 1930s," *Journal of the Society for the Study of Architecture in Canada* 36, no. 2 (2011): 5-12.

[23] Mary Vipond, *Listening In: The First Decade of Canadian Broadcasting 1922-1932* (Montreal & Kingston: McGill-Queen's University Press, 1992), 174-175.

[24] Anne F. MacLennan, "Circumstances Beyond Our Control: Canadian Radio Program Schedule Evolution during the 1930s" (PhD diss., Concordia University, 2001): 98-145.

[25] Westinghouse Blue Ribbon Air Pilot Radio. "Splitting Hairs between Toronto and Cincinnati,".*Maclean's Magazine*. 49 (October 1, 1936): 3.

[26] *Canadian Broadcasting Corporation Annual Report, 1946-1947* (Ottawa: King's Printer and Controller of Stationary, 1947), 31.

Front cover detail, Eaton's *Radio Catalogue*. 1927-1928.

Selling Radio:
How Radio Manufacturers Designed Desire

2

The challenge of marketing and advertising a new, previously unnecessary product is overcome once its unique possibilities become evident. Stewart Ewen argues that "advertising plays a conspicuous and powerful role in people's lives. It pushes a vision of life which says that satisfaction is available across the retail countertop."[1] Advertisers were tasked with creating a 'vision' of radio in Canadian households, despite the fact that radio's qualities were anything but visual. Magazine and newspaper readers were unable to hear radio for themselves, so manufacturers had to create a compelling vision of radio through image and text. To promote this new and unfamiliar technology in the 1920s and 1930s, advertisers had to explain radio's technological merits, develop an audience for radio programming, and demonstrate that this electronic device could find a place in the home. Even after radio became widely adopted, it continued to be sold by promoting its technology and design, and promising the listener access to a world of sophistication. Advertising helped to design radio culture in Canada, heralding the new medium in the pages of Canadian newspapers.

Advertisements helped Canadian consumers visualize the dream and wonder of radio in their homes. As many Canadians purchased radios, the firmly-ingrained memories of the designs and models attest to the manufacturers' success in branding their product as a necessary part of the modern home.[2] The challenge was to communicate the impact of radio in print.

Starting in 1922, Canadians were required to license radio receiving sets. In a year, Canada had 9,956 registered radio owners. This tripled by 1924 (to 31,609) and again in 1925 (to 91,996).[3] By 1930, the number of registered owners quadrupled to 414,146 and continued to increase steadily over the next decades, exceeding 2.2 million in 1953, the last year Canada collected the license fee.[4] The explosive growth in radio ownership fueled and was fueled by the radio advertising and manufacturing industries.

In the early twentieth century, radio emerged as one of many new consumer household products that included mechanical refrigerators and electric vacuum cleaners. But the purchase of radios far outstripped that of other new products.[5] In rural Canada, a lack of electrification contributed to low demand for new electrical appliances. However, since radio could operate with batteries or electric power, they became popular throughout Canada. As Colin Campbell suggests, "modern consumers will desire a novel rather than a familiar product, largely because they believe its acquisition and use can supply them with pleasurable experiences that they have not so far encountered in reality."[6] Radio's novelty contributed to its rapid rise and this demanded innovative approaches to its advertising.

Advertisements framed the Canadian popular imagination of radio prior to its entry into the home. They also helped advance radio's use, evolution, and design. Although radios were first popularized by hobbyists who made their own crystal sets, early radios were expensive floor models and their purchase was limited to the few who could afford them. Print advertising guided most Canadians through the quickly-evolving technology, as it transformed into tabletop models and became smaller and more affordable. Advertising helped to shape the design of radio receivers, listening practices, and the place of radio in the home.

Early advertisers cast a wide net when promoting radios. Roland Marchand describes these advertisements in part as "reflective of the 'reality' of the social aspirations of American consumers ... more accurately described as 'social fantasy' than 'social reality'."[7] Some early radio advertising brought every feature to the reader's attention: radio's technical capabilities, its impressiveness in the home, its associations with high culture, and the wealth of programming available to listeners at the turn of a knob.

The 1922 Marconi advertisement discussed in the first chapter (Fig. 1.1) provides an excellent early example of this strategy,

> Radio! A new gift and a permanent one. A daily reminder of your thoughtfulness, it is unique in that it furnishes the home with constant contact with the world and its doings. But remember — there's nothing 'just as good' as the Marconi Long Range Receiver.[8]

The text of the advertisement is encircled by the myriad of programming that the long-range receiver can bring to the listener: orchestral music, a jazz band, a hockey game, storytelling, and political news.[9] Campbell describes this as the "basic motivation underlying consumerism ... the desire to experience in reality that pleasurable experience the consumer has already enjoyed imaginatively."[10] This early advertisement fed the imaginations of the buying public with radio's possibilities. Printed in the first year of Canadian commercial broadcasting, as the radio was still emerging as a valued household appliance, it demonstrates all that radio has to offer.

Radio advertising took shape in the 1920s and 1930s by employing multiple strategies. These simultaneous and overlapping approaches included technology-based radio advertising that appealed to radio enthusiasts eager for innovations and new features, listener-based radio-advertising that fed the imaginations of people drawn to the new medium, and finally, aesthetically-based radio advertising that appealed to people's sense of identity and pride, especially in the home These strategies were revisited in postwar radio advertising to keep up with technical innovations and new design ideas.

Technology-based radio advertising

While some advertisements were broadly targeted at any consumer, most were aimed at specific groups. The First World War delayed the start of commercial radio in Canada, but Canada's CFCF in Montreal may have been the world's first radio station, putting Canada at the forefront of broadcasting.[11] Amateur broadcasters had experimented with radio before the war (CFCF's first license was issued to XWA in the 1914-15 fiscal year), but test programming started in 1919, immediately followed by the arrival of a Marconi Company store to sell radios in Montreal.[12] Radio began in the hands of the hobbyists, some of whom became engineers for early broadcasting stations, applying their military training acquired during the war. Radio's early development rested on a 'do-it-yourself' foundation as young men and boys assembled their own crystal sets, eagerly seeking new stations.[13] This amateur broadcasting preceded commercial radio, which was not established until 1922 when commercial licenses first became available. The early adopters of radio technology did not have to be convinced of the merits of broadcasting, but the general public would need to be sold on the benefits of radio if they were going to buy a radio set. In the 1922 book *How to Retail Radio*, radio merchandisers were advised, "don't let a salesman dodge the issue with generalities when a confirmed 'radio nut' comes in. Better let him turn the prospect over to someone who knows."[14] The superior knowledge of expert potential buyers forced advertisers to extol the technical virtues of the product for some consumers, while creating an imagined space for others.

Northern Electric was an early and active member of the radio community. In its advertisement from the pages of *MacLean's Magazine* in 1924, readers were summoned to "Tune in with the world of Entertainment!" (Fig. 2.1).[15] The advertisement is directed at a wide variety of potential consumers. While the text and images focus on radio's

Fig 2.1

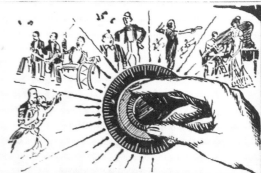

Tune in with the world of Entertainment!

The new Northern Electric Set R-11 will give you full, genuine Radio enjoyment.

While you delay—while you doubt—every day, every night, all over the world, bands are playing for you. You are missing wonderful entertainment that an R-11 set would bring to your home. Singers, lecturers, comedians, dance orchestras are yours to command.

Will you close your ears to this world of enjoyment?

Write to us for information about radio. Consult our radio engineers free of charge.

See a *Northern Electric Dealer.* Ask him to let you 'listen in' on a R-11 set—the latest development in small receiving apparatus.

R-11. A Northern Electric Product.

Handsome mahogany cabinet, compact and light, with one Peanut Tube.

For ideal results order R-11 with R-15 Amplifier and two Peanut Tubes.

Northern Electric
COMPANY LIMITED
The people who made your 'Phone

technical aspects, the headline extols radio's programming options. A manicured hand at the tuning knob hints that the listening audience would include women. The choices start with a dancing couple in evening wear, followed by formally-dressed musicians, two men in conversation (perhaps comedic), a symphonic conductor, and finally a coloratura soprano and her accompanist at a grand piano. This world of entertainment is dominated by music and an imagined access to high culture that musicians provide. The prominence of the dial at the top of the advertisement speaks to ease of use — unlike the early crystal sets, a dial made it easy for anyone in the home to 'tune in' to a wide variety of programming. Yet, the remainder of the advertisement is mindful of the technology-focused experts who were the earliest radio enthusiasts.

The design of the radio and its place in the home does not figure in this advertisement. Aside from the reference to "singers, lecturers, comedians, and dance orchestras" the text focuses on radio's technical aspects. The receiving set, an R-11, is technical enough in 1924 that under the earphones and the command "Listen!" readers are instructed that they can consult with Northern Electric "radio engineers free of charge" by writing away for the information. While the potential listener is told that "you are missing wonderful entertainment that an R-11 set would bring to your home", the small print tells the reader that the mahogany cabinet comes with one Peanut Tube, but an R-15 Amplifier and two Peanut Tubes are recommended. The apparatus depicted looks technical and has not been designed for display in a parlour. A further appeal to this radio's technical superiority comes with the tagline beneath Northern Electric Company Limited – "The people who made your 'phone'"– with a list of branches in twelve cities across Canada. Finally, the advertisement lists the company as the owner and operator of CHYC, a Montreal radio station that acquired its license in 1923. The technical expertise of the company is certainly evident in the advertisement. Competence was a concern for early purchasers looking for a reliable product — particularly those paying 'top dollar' to place this new invention in their home, and also for the enthusiasts who could assemble radios on their own but sought the sophistication of a technically well-made radio from a reliable manufacturer.

Technical claims heralded the entrance of the radio into the home. "No longer is it necessary to have unsightly batteries, ground connections or aerials, in order to enjoy radio reception," states an advertisement from 1924 (Fig. 2.2).[16] This advertisement shows the radio on a piece of furniture beside a vase of flowers. A special looped antenna is concealed in the set, as an inducement to purchasers who might be more concerned with the appearance of their home than the quality of their radio reception. The advertisement remarks on its features: "The six Radiotrons are operated by dry

Fig 2.1 Advertisement for Northern Electric Company Limited. *MacLean's Magazine*. 1924.

Fig 2.2 Advertisement for Canadian General Electric Co. Limited. *Montreal Gazette*. 1924

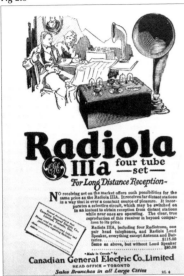

Fig 2.3

cells and the set may be carried with perfect ease. The cabinet is beautifully finished mahogany, equipped with a leather handle for carrying from place to place."[17] Its price of $350 put this set out of the reach of an overwhelming majority of Canadians in 1924. For the economy-minded, an alternate set lacking batteries and antenna sold for $115.[18] The Radiola was manufactured in both Canada and the United States. For consumers more focused on quality of reception, the advertisement indicates that an external loop antenna would provide additional range and that further information would be available on request with an "illustrated booklet and particulars regarding Radiola."[19] Such technical booklets appealed both to the hobbyist and the uninitiated wanting to learn more about radio.

Canadian General's Radiola IIIa was specifically advertised as a set "for long distance reception" (Fig. 2.3).[20] The technology is prominently displayed in the foreground of this ad with array of knobs and tubes.

> No receiving set on the market offers such possibilities for the same price as the Radiola IIIA. It receives far distant stations in a way that is ever a constant source of pleasure. It incorporates a selective circuit, which may be switched on in an instant to obtain reception from distant stations while near ones are operating.[21]

In the background, two men in formal dress listen together. The long distance reception certainly refers to the hobby of DXing, popular among radio aficionados.[22] As noted by Lizbeth Cohen, "Even workers who bought increasingly affordable, ready-made radios spent evenings bent over their dial boards, working to get 'the utmost possible DX' (distance), and then recording their triumphs in a radio log."[23] While Cohen here addresses grassroots urban cultures, DXing was equally popular in rural communities and nationwide in Canada. Reynold Wik explains that "After buying a radio, the new owners became obsessed with efforts to reach out as far as possible. Apparently there was a fascination with the notion that now you could annihilate distance. This urge became so strong that *Radio News* in 1923 called this impulse 'the radio itch'."[24]
This hobby or obsession continued into the 1920s and 1930s, especially in the late night hours after regular family programming. Advertisements featuring long distance reception signalled the transition to pre-assembled radio receiving sets.

Concerns over the appearance of the radio in the home are dispelled by a 1927 advertisement, which states "the new A.C. Stromberg-Carlsons are utterly free from the need of attention — always ready to be placed in use at the turn of a single switch."[25] A well-designed cabinet connected to a wall plug is prominently displayed, while an array of batteries is crossed out. Models ranged in price from $20 to $565, but the simplicity of the Stromberg-Carlson A/C receiver was primary. The radio's ease of operation and its

Fig 2.4

harmony of design with interiors became a repeated advertising appeal.

De Forest & Crosley announced further refinements to the "new radio . . . built in Canada to meet Canadian conditions".[26] Angels trumpeting the news from the clouds suggested the importance of this development. The advertisement announces that "DeForest & Crosley presents a complete new line of radio receiving sets, each model designed expressly to meet the long-distance requirements of Canadian conditions. . . . It has developed circuits which in efficiency, in simplicity and tone quality represent a distinct Canadian achievement . . . renew the miracle of radio . . . giving the most experienced radio fan a thrill of surprise and delight."[27]

The innovation and adaptation to Canada (providing radio reception over long distances) was the first selling point, but the myriad of choices, price point (from $22 to $395), and technical updates paralleled the evolution of radio in the United States. The flourishing industry in Canada was reflected in the rising sales and the increasing availability of radio broadcasting.

A Northern Electric advertisement from 1924 proclaims radio as "The Crowning Achievement of a Scientific Century" (Fig. 2.4).[28] According to Artemis Yagou, this "early domestic type [of radio] reveals the iconographic influence of . . . scientific instruments and is indicative of the still vague conception of the new medium by the public."[29] In his book *Objects of Desire: Design and Society Since 1750*, Adrian Forty notes that "the [British] buying public was unconcerned about what the receivers looked like and was more interested in how many and what type of valves each model had."[30] This

Fig 2.3 Advertisement for Canadian General Electric Co. Limited. *MacLean's Magazine*. 1924.

Fig 2.4 Northern Electric advertisement. *MacLean's Magazine*. 1924

advertisement appeals to the scientist and the technician. Judith Williamson draws on Lévi-Strauss in *Decoding Advertisements: Ideology and Meaning in Advertising* to reinforce the impact of science on the "relationship between nature and culture definable in terms of this particular period and civilization."[31] Such an ongoing dialogue with science is embodied in early radio advertising. Northern Electric's advertisement visually positions the radio between the 'ghosts' of the automobile and the telephone, modernity's other achievements. The telephone was also another Northern Electric product, further underscoring the company's scientific authority and technical expertise. The text of the advertisement promotes the R-4 as an instrument of "perfection . . . absolute precision, the accuracy . . . all the marvelous clearness, modulation and richness of tones . . . this supreme achievement . . . a beautiful instrument".[32] Toward the end of the advertisement, the reader is told of a "handsome mahogany cabinet of genuine beauty and symmetry" and that this was the "set used by H.R.H. the Prince of Wales at the E.P. Ranch near Calgary".[33] Radio's technical and scientific attributes were its primary selling points in many advertising campaigns in the early years.

In these three early Canadian advertisements, the consumer is largely absent. The advertising copy is focused on radio's technical aspects. The knobs, dials, antennae, batteries, and other parts of the radio appealed to the enthusiast with radio experience, and the wealthy consumer who could afford the latest and best inventions for their homes. But while the technical merits of the radio would remain important, the scientific elements were muted after the first few years of commercial broadcasting. Early radio advertising that focused on the tubes and inner workings of the radio was short-lived. In the same way, the workings of the radio started to be disguised so that their technical aspects, which had great appeal to enthusiasts, became secondary to the needs and desires of other listeners in the home.[34]

Listener-based radio advertising

Once the technical credentials of the radio were established, advertisers encouraged people to imagine the auditory experiences they could have as radio listeners, and communities of which this would make them a part.[35] An Atwater Kent Radio advertisement from 1925 rhapsodizes,

> what a wonderful, delightful difference it makes – just think what it means in one evening the thrills of a lifetime are crowded into a few short hours. Set the dials and the melodies of a famous orchestra flood your home, another touch and you hear a lecture from miles away — turn again and you have the news of the day or the sweet voice of a renowned singer generously broadcasting for your entertainment.[36]

The image shows a woman in a stylish gown languorously draped across a chair while listening to the radio, conveying sophistication and cultural capital. The radio looks like a phonograph and a table flanked with candlesticks suggests a grand piano in the background. Through radio, listeners could imagine themselves experiencing the lifestyle of the wealthy, no matter how modest their own home. Roland Marchand described similar advertisements in which "the ladies, more emotional but nevertheless refined and decorous, draped themselves gracefully over the huge chairs and divans, staring pensively into space."[37] This luxury and grace connecting the listener and radio marked the early stage of women as ornaments for technology in advertising. Such images of wealth would eventually be brought to working-class and middle-class families, as predicted in *How to Retail Radio*,

> Radio sets will ultimately be so reduced in price [to] be sold to people of moderate means. The installment plan of sale will place them in millions of homes just as it has placed millions of phonographs and household appliances. What was the rich man's toy will become the workingman's pleasure.[38]

Anticipation of this cultural richness was a selling point for Stewart-Warner's Matched-Unit Radio. Described as "The Radio You Have Been Waiting For", the radio set is placed in the home with a smiling mother and her daughter with her doll. The advertisement describes the radio as a "companion for the entire family," emphasizing features to interest everyone.[39] The technical aspects of radio were selling points: "Full, clear tones retain their natural timbre. Abundant volume fulfills every desire. Selectivity makes distance reception a pleasure,"[40] as is ease of use with the new "special feature – a wave length dial that eliminates the necessity of a log and enables you to tune in the stations directly from the wave lengths listed in the radio programs."[41] According to Susan Smulyan, "Because engineers could not discover a way to improve long-distance reception, new radio receivers focused on improved reception quality and helped create a new type of radio fan."[42]

Radio's audience expanded beyond the hobbyists and enthusiasts (who focused on distance) to include the listener who enjoyed a higher quality of sound when tuned into local programming. A column on the right shows how the "Matched-Unit" could be paired with a variety of cabinets to fit into the home. The text echoed this message, "Every model is beautiful in design and workmanship. The tuning controls are conveniently arranged and blend perfectly with the unusually fine walnut finish. The complete unit adds taste and richness to your home surroundings."[43] Since radio was still a luxury item and it occupied a great deal of space in the parlour, attention to design was a selling point.

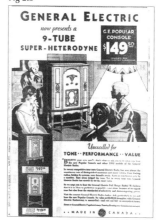

Fig 2.5

Fig 2.5 Advertisement for General Electric Co. *Winnipeg Free Press*. 1931.

Fig 2.6

In the same spirit, announcing its 1932 models, General Electric asked readers to "Believe your own ears" (Fig. 2.5).⁴⁴ In the advertisement, a family is gathered around a large console, the father in a dinner jacket and the mother in a sleeveless evening gown, indicating anticipation for the event of listening to the radio in their home. Three styles of console provide a focal point for the family 'listening in'.

Although technical features and ease of use continued to be key elements of radio advertising, the connection to the world and the imagination were essential to bring all family members into the radio-enjoying audience. As early as 1922, a Northern Electric advertisement proclaimed that "The Children too can Enjoy Radio" (Fig. 2.6).⁴⁵ Although children in most homes were not allowed to touch the dial on their own, a small portion of daily programming was directed toward children. The low production values of an 'Auntie' or 'Uncle' program permitted the genre of children's programming to flourish in Canada, maintaining a position in the schedule in the early decades of broadcasting.⁴⁶ These programs usually included songs and stories from fairytales or comics read from the newspaper. They were a staple for most radio stations in Canada and the United States and were broadcast during children's bedtimes. As Noah Arceneaux explains,

> Many stations . . . broadcast bedtime stories and department stores had a particular motive . . . If a station could encourage a child to regularly listen to its bedtime stories, the station was simultaneously establishing a relationship with the primary care-giver, most often the mother.⁴⁷

The integration of the radio and its programming into the daily routine of the home was an important element in marketing radio.

In Northern Electric's advertisement, the radio sits on a table as three children listen in with earphones: a little girl in an armchair, her brother at the table, and a third perched on an ottoman. Their shared enjoyment floats above their heads, with images of Mother Goose, a pirate ship, Puss 'n' Boots, a clown, and a dog. As the children sit in rapt attention, the advertisement reads,

> Don't you remember those bed time stories of princes and pirates, adventures in Fairyland. Lack of time perhaps prevents you from reading aloud to your children. Why not let RADIO do it? Experienced story tellers, persons who understand fairies and goblins, who know the haunts of Wendy and Peter Pan will call forth nightly those idols of Shadowland to delight the little folk. Education in the romances dear to childhood is as important as the multiplication table.⁴⁸

The advertisement speaks to the convenience of radio for busy parents relieved of the task of reading to their children. Education is called into play as well, to justify the purchase of a radio which everyone in the family can enjoy. The copy makes no mention of the technical merits of the radio receiver. It does not even mention a model. The appeal is based solely on access to children's programming and the support it will provide for parents.

The Eaton's *Radio Catalogue* extended the imagined radio audience even further to include grandparents. The 1927-28 catalogue depicts a white-haired gentleman reclining in an upholstered chair in front of a floor model radio in a richly-carved cabinet (Plate 2.1).[49] He sits facing the set, solving a newspaper crossword as he listens to the radio. French doors open up to a vista of his property. In the distance and set in the clouds, we see performers at microphones demonstrating the variety of program choices offered on the airwaves. His relaxed manner as he listens to the radio leisurely demonstrates how seamlessly radio integrated itself into the daily life of its listeners.

Richard Butch notes a similar development in *The Wireless Age*, as it transformed from a hobbyist magazine to one with a listener-based readership.

> Covers now featured color drawings of people listening, Norman Rockwell-style romanticizations of upper middle-class American domestic life. The May cover depicted a well-dressed woman sitting and listening with headphones in her parlor, the radio on a table with wires and batteries hidden in furniture cabinet, and playing along with the music with her hands.[50]

Eaton's *Radio Catalogue* of the previous year also featured a white-haired gentleman, this one at a kitchen table (Plate 2.2).[51] Here, the radio is a table model but also easy to operate, as we see the husband adjusting the knob while looking at his wife as she dries the dishes. The cover suggests that the radio has found its place in the home, fitting into the rhythm of daily life.

One of the most common strategies for advertising radio in print media was to depict performers rising out of the set, genie-like, and into the home. A 1926 advertisement for Canadian General Electric's Radiola shows a larger-than-life violinist, emerging from the radio receiver (Fig. 2.7).[52] While he is undoubtedly imagined by the stylishly-coiffed women listening to the left, the tagline is "The Product of Radio's Master Minds."[53] The visual aspects of this advertisement are directed to appeal to the female reader, yet in contrast the text focuses exclusively on the radio as a technical marvel, "the reward of ceaseless research and experiment on the part of Radiola engineers – master minds who have robbed radio of its mystery and reduced radio-reception to utmost simplicity."[54]

Sparton Radio's advertisements also feature images that emerge from the radio and in this case, overwhelm it. "Face-to-Face Realism: Big Game Reception that is truly ALIVE" is illustrated with figures of three football players complete with shadows in the

Fig 2.6 Advertisement for Northern Electric Company Limited. *Montreal Gazette*. 1923.

Fig 2.7

Fig 2.8

Fig 2.9

Fig 2.10

background to intensify the realism (Fig. 2.8).[55] The athletes dwarf the traditional floor model set in a cabinet. Ray Barfield explains in *Listening to Radio, 1920-1950*,

> As the 1920's began to wane . . . radio 'magic' spread to . . . remote places. . . . Owen Lyons, Jr. recalls . . . my dad was an avid fan of the St. Louis Baseball Cardinals. He bought an Atwater Kent console radio and two telephone poles . . . ran a wire between the two poles, with a connecting wire down to the window near the radio.[56]

Lyons's memory points to a major change wrought by radio. In Canada, sports had been limited to live events and newspaper summaries, however "once radio introduced the play-by-play over the air, the sports fan base and its participation was irrevocably changed."[57] The advertisement draws on the excitement of this change, exclaiming that,

> with FACE-TO-FACE REALISM you . . . sit beside your announcer . . . to see as he sees and to feel as he feels. . . . The seething excitement gets into your blood. The roar of the crowd as it surges to its feet carries you with it. Breath-taking moments for spectators hold the same thrills for you. . . . This is an invitation . . . for you to HEAR and FEEL this remarkable radio development.[58]

The feel of the game, its movement and the engagement of the senses represented a radical change that created a fan base for mediatized sports. As a 'hot' medium, radio more readily captured the movement of sport than could a newspaper. It extended its reach simultaneously in the home and the arena. Similar home audiences emerged for the orchestras, symphonies, comedies, and other radio broadcast events.

The Face-To-Face Realism campaign continued, revealing another facet of the radio audience as imagined and depicted in radio advertising. "Face-To-Face Realism: Radio music that exalts – and thrills" features organ music, a popular performing art in Canadian cities (Fig. 2.9).[59] But organ music is also emblematic of the religious programming available over the airwaves.[60]

> Radio stations across the country took advantage of the eagerness of the churches in their cities to deliver their messages to a broader community. Church services constituted a dominant daytime element of Sunday broadcasts. Using local churches, Sunday services were set up as 'remotes' taking advantage of the immediacy of the medium.[61]

The religious associations are hinted at in the copy: "music that exalts. . . . You will find every exquisite shade of light and shadow that the master puts into his music. You will seem to be conscious of his living presence."[62] While sometimes controversial, as when a variety of religious services were broadcast by the same station or when one denomination criticized another, the radio provided a way to extend religion into the

Fig 2.7 Canadian General Electric Co advertisement. *Maclean's*. 1926.

Fig 2.8 Sparton Radio advertisment. *Montreal Gazette*. 1949.

Fig 2.9 Sparton Radio advertisement. *Montreal Gazette*. 1929.

Fig 2.10 Advertisement for Layton Bros. Limited. *Montreal Gazette*. 1936.

home in the form of live religious services, sermons, hymns, or music.⁶³ As a symbol of the church in image and sound, the pipe organ was well-suited for radio advertisement.

The advertisements of the 1920s are also full of offers for further information or, as in this case, an invitation to take the radio home to test it, without obligation to purchase. In the foreground, to the right of the text, Model 301 is displayed in a large, imposing cabinet sharing some of the qualities of the impressive pipe organ that dwarfs it in size. The larger-than-life imagery in this advertising campaign was widely used by other radio manufacturers to communicate the sense of a product that was larger than the receiver or its cabinet. Radio was larger than life. It connected to the world beyond.

The Layton Brothers "Magic Voice of an RCA Victor Radio" prominently featured the candidates for the August 17, 1936 Quebec provincial election: Adelard Godbout and Maurice Duplessis (Fig. 2.10). The focus was the Model 9K-1, with "a radio development exclusive to RCA Victor receivers. Magic Voice is radio speaking in a clear, new voice as if the speaker were right in your very living room."⁶⁴ The advertisement also included the Mode 5T-3 for a little over a third of the price of the floor model. Recognizing that many readers already owned radios and the financial strains of the Great Depression, the advertisement offered "generous trade-in allowance on your obsolete radio" as an incentive to upgrade. Readers were no longer enticed with offers of technical information on request. Instead, they are invited to hear the difference for themselves at a nearby store. The 1936 election was an opportunity to sell radios to listeners disappointed by poor reception and eager to follow the campaign in real time. Released only three days before the election, the advertisement took advantage of this enthusiasm. Radio coverage of elections was as appealing as play-by-play coverage of sports. There was no need to wait a day for the newspaper summary of the election results. Radio allowed Canadian electors to get the news as soon as the candidates did. While this advertisement stressed the importance of audio quality for listening to the debate, its connection to the 1936 election departs from the standard fare. Connection to the larger world is a repeated trope employed in radio advertising.⁶⁵

Aesthetically based radio advertising

Radio's entry into the daily lives of Canadians was established first on the technical merits of the pre-assembled radio receiver set, and, secondly through the flights of fancy demonstrated by the possibilities of programming and connection. But the last stage was to gain entry into the home as a 'part of the furniture', eliminating any sense of experiment or threat to family life.⁶⁶ Jean Baudrillard posits that

> training in systematic, organized consumption is the equivalent and extension, in

the twentieth century, of the great nineteenth-century-long process of the training of rural populations for industrial work. . . . This process reaches its culmination in the twentieth century in that of consumption.[67]

The sense of self as a consumer begins to supplant that as a worker, thereby allowing the worth and identity of a person to be assessed by the value and accumulation of their purchases. It is in this process of conversion to consumers that the listening audience is fully incorporated as purchasers of radios. The first stage of this broadening of the radio-buying public was established with the lure of the larger world of programming, but in the second stage, the radio had to fit into the household.

Gaining entry into the home meant a change in appearance and a smoother, less messy and leaky function. William Boddy explains that retailers were advised this transition would be eased by changing their targeted consumers.[68] For example, from *Radio Dealer* (1923): "And, by all means, don't talk circuits. Don't talk in electrical terms . . . You must convince every one of these prospects that radio will fit into the well appointed home."[69] This is a completely different approach than that used to woo to the technically savvy radiophile. Instead, it sells the beauty of the cabinet in the home of the consumer. Here, the gendered and class aspects of radio consumption become evident. Michael Brown and Corley Dennison chart the changing attitudes toward radio, analyzing a *Radio News* contest of 1926 which aimed to identify consumers' preferred designs for home receivers.[70] Ten percent of the entrants were women. The editor's wife, Mrs. Hugo Gernsback, who is credited with the idea for the contest, asserts that "the lady of the house must be pleased first [with a] piece of furniture that will harmonize with other furniture in the room."[71] At this time, Brown and Dennison argue, the technical distinctions between different radio models became less pronounced and thus required promotion through "aesthetics rather than technical superiority". In other words, the house-proud could also drive a market for radio.[72]

The 1930-31 Eaton's *Radio Catalogue* speaks to the transition of radio technology into the home (Plate 2.3).[73] The cover is dominated by the globe of the Earth encircled by a bolt of electricity, evoking power, connectivity, and the magic of radio.[74] The globe in a starry night sky shines down on a home radiating electric light and with an aerial on its roof. The connections to the globe, electricity and modernity are clear. The inset depicts a well-dressed couple in their well-appointed living room with a few indicators of their status (a clock on the mantle and a painting on the wall), but focuses on the husband's efforts to tune into a radio station. The pattern of the radio console echoes the upholstery of the armchair, knitting the set into the home. This advertisement shows the world brought into the home by radio.[75]

Fig 2.11

Fig 2.12

While the gendered images of radio consumer are clear in the definition of the public and private spheres, male technical superiority is not always asserted, because ease of use is essential to making a successful pitch for radio in the home. So while Northern Electric showed a well-manicured female hand turning the dial in 1924, the *Eaton's Radio Catalogue* of 1930-31 shows the husband entrusted with the dial. While control of the dial may alternate between husband and wife in advertisements, they are absent when the advertising focuses on radio's technical merits or the beauty of its cabinet design in the home. Advertisements stressing radio's programming content usually portrayed the family enjoying their radio.

The visual hyperbole of radio's larger-than-life programming figured in "The Always a Marvel Now a Giant for Power," a 1925 advertisement for Northern Electric. (Fig. 2.11) In this ad, radio is lifted high by a figure who appears to be the Mars – the Roman god of war. The power of the set is emphasized by repetition:

Power, that brings in stations you never heard before.

Power, that 'clears up' stations in the 'dim areas'.

Power, without harshness.

That's what Northern Electric Sets now offer, because of the Giant Power in the new Peanut Tube.

It's power within that makes them win.

Ask the Northern Electric Dealer to show you these powerful sets.[76]

These radios were sold without a cabinet and priced starting at $30. While the advertisement effectively associated 'radio' and 'power', such a set might appeal on its technical merits yet not attract consumers looking to buy a set for their living room. The move away from this demonstration of technical power and innovation is introduced with the concepts of beauty, style and, craftsmanship.

The fascination with radio's larger-than-life entertainment continued into the 1930s. In 1932, Northern Electric's advertisement "These NEW Sets will Amaze You!" shows a statuette positioned atop the "Miss Concerto" set and a full-scale showgirl emerging from the larger "Miss Symphony" set (Fig. 2.12).[77] The image of beauty continues to their naming: Model 101 "Miss Concerto", Model 120 "Miss Symphony", Model 80 "Miss Tango", and Model 60 "Miss Rumba". This naming may have appealed to both men and women, but its most significant aspect is the dual appeal to classical and popular dance music. At this time, the variety of radio sets multiplied with the introduction of tabletop models. While the earliest radios were often placed on a table or hidden in a cabinet, these new sets were advertised for their "astounding advances in radio construction and performance", as "exceptionally beautiful examples of the cabinet makers' art" and also as evidence of Northern Electric's reputation as "pioneers and leaders in the development of electrical sound equipment".[78]

Craftsmanship, beauty, and design continued to be an important element of radio advertisement.

Advertisers highlighted the contrast between old and new. One manufacturer sought middle- and upper-class consumers ready for 'a new conception of radio' in 1925. . . . With this new receiver, the advertisement exulted, art had triumphed over 'mere mechanics'[79]

In 1937, *Chatelaine* featured an advertisement for the Westinghouse World Cruiser Radio under the headline "Only Master Craftsmanship Can Infuse Wood and Metal with this Living Beauty" (Plate 2.4).[80] The advertisement depicts a colonial-era craftsman dressed in white wig, breeches, hose, and jacket with ruffled cuffs hard at work on a Chippendale chair. Here, the artisan represents craftsmanship and quality. The console design manages to look both modern, with an Art Deco quality, and also historic, with its symmetry and fine woodwork.[81] The accompanying text tells us, "It takes more than four legs, a back and a seat to make a Chippendale chair. It takes more than wood and metal, more than features and 'gadgets', to make a Westinghouse Radio." Even in the depth of the Great Depression, radio remains a well-crafted thing of beauty sought after for its perfection. Its traditional design, like the Chippendale chair, fits into the home, conveying "outward beauty worthy of the finest tradition of the cabinetmakers [... and] satisfying performance which provides further evidence of Westinghouse engineering leadership."[82] Here, the prestige of a cabinet radio was still significant despite the availability of tabletop models, indicating the position that radio had taken in the home.

The traditional cabinet had its roots in the first years of radio's production and advertising. In 1926 "The Merriest Christmas You've Ever Known" announced the DeForest Crosley Radio (Fig. 2.13).[83] Radios were most frequently advertised in December; as a large and expensive purchase, they were considered excellent gifts. The text of the advertisement reads,

Let radio make this the merriest Christmas you've ever known. Let all the good cheer filling the air come right to your fireside. Every detail of DeForest & Crosley radio is expressly engineered to meet Canadian conditions.[84]

Aside from the text, the only image in the advertisement is the radio cabinet. The shine of its top, the intricate carving of its panels, and the traditional styling of its legs are featured. The text further extolls its qualities and performance,

The Drum Control . . . selects programs with unerring accuracy. The Shielded

Fig 2.13

Fig 2.11 Northern Electric advertisement from Layton Bros Limited. *Montreal Gazette*. 1925

Fig 2.12 Northern Electric Radio advertisement. *Montreal Gazette*. 1932.

Fig 2.13 DeForest Crosley Radio advertisement. *Montreal Gazette*. 1926.

> Chassis . . . shuts out local interference and unwanted stations. The Wheatstone Bridge principle of balance . . . eliminates squealing or howling. The Reserve Power necessary to overcome the distances . . . found in Canada. The Special D.C. Tone Chamber . . . recreates every note with life-like exactness.[85]

Here, performance intertwined with beauty provides the promise of Christmas happiness.

The aesthetics of taste and the association with elite surroundings played a strong role in selling radios as a desirable possession.[86] Stromberg-Carlson's 1931 advertisement "In keeping with the best in your home" tells readers that:

> Pride in your home need not be compromised in purchasing a radio. Matching the best in furniture your home contains with a Stromberg-Carlson costs but a few dollars more in down payment. Letting people see and hear that there is no radio finer than the one you possess means only the equivalent of a month or two more of installments that you would pay on an ordinary receiver.[87]

The headline, on optical centre, takes the eye immediately above to an ornate chandelier. Decorative crown mouldings lead the eye across drapes and to a pilaster, which bring it back to the Stromberg-Carlson receiver, an ornate well-carved piece of furniture 'at home' in its opulent surroundings.

> Being able to say you own a Stromberg-Carlson is itself a satisfaction. For everyone knows the tonal quality of a Stromberg-Carlson is beyond musical criticism. Its size has not been reduced to save materials with consequent impairment of tonal range. Its parts have not been lightened or cheapened to compete with receivers sold only by their price. Its cabinet and its carved decorations have not been made of substitutes for genuine walnut. To keep up the quality of your home in radio as in all other particulars: - 'There is nothing finer than a Stromberg-Carlson.'

Stromberg-Carlson pitched quality, its refusal to compromise on the highest standards. Price reflected their market placement, with lower priced receivers ranging from $199 to $465 and their Multi-Record Radio (which combined the radio and phonograph) retailed for $845, well beyond the budgets of most consumer. As most homes acquired a radio, some radios remained aspirational purchases, out of reach but admired. While this advertisement extolled the product quality, it provided little information about the radio's performance or technical features — quality and an unwillingness to compromise was the message.

Stromberg-Carlson had been promoting up-market radio for some time. A 1929 advertisement described their radios as "Truly works of art".[88] The advertisement features another artisan, crafting a fine violin in a Renaissance atelier, with a Stromberg-Carlson displayed prominently in the foreground.

> There is much about every product that reflects the people and plant back of it. Just as the fine craftsmanship of di Salo and the Guarnari set their instruments apart from all others of their time, so, too, the painstaking skill of Stromberg-Carlson has distinguished its receivers from other radio instruments of today.

By comparing the manufacture of their radio to the workmanship of a sixteenth century Italian violin maker here Stromberg-Carlson evokes a lasting tradition. The analogy extends to sound quality, "Stromberg-Carlson engineers match distinguishing exterior charm with an interior precision resulting in tonal beauty truly distinct from any set you've ever heard."[89] According to Michael Brown and Corley Dennison,

> The massive, elaborate radio sets allowed wealthy listeners to purchase a radio set that reflected their status. . . . Owners of these sets shared a common technology, but their radios were housed in cabinets that set them apart and identified their status.[90]

The Stromberg-Carlson campaigns appealed to such distinctions and the trapping of the upper class. Once enthusiasts ceased to be the dominant group of radio purchasers, cabinet design and the radio's place in the home became increasingly important to advertisers.

Fada Radio's advertisement of 1925 depicts the radio in a beautiful, classically decorated home. The company's tagline was "The Grand Piano of the Radio World", a claim borne out by one look at the advertisement (Fig. 2.14). Fada radios were sold for appearance, assuring readers that technical innovation was unlikely for years, making the investment in a well-designed cabinet worthwhile. The copy proclaims "Cabinets that harmonize with beautiful interiors",[91] which is further illustrated by a Fada console in an opulent foyer complete with a curved staircase, French doors, a grandfather clock, Persian rug, and a majestic window framed by drapes.

> Just as the grand piano embodies the highest ideals of musical and decorative art, so in the world of radio, Fada leadership is established beyond question. Fada Radios are masterpieces of the cabinet maker's art. A fine range of beautiful models invites you to choose the design that will harmonize with the individual furnishings of your home.[92]

The advertisement is pitched to consumers wanting the perfect adornment for their home, while dispelling the worries that such a big investment might become obsolete. "You may purchase a Fada with reasonable assurance that no essential changes in radio invention for years to come will render it obsolete or less desirable." Here, the consumer is encouraged to value taste and beauty over performance or innovation.

While the home, as part of the private sphere, was considered a female domain in the early twentieth century, many radio advertisements focused on the instrument rather

Fig 2.14 Advertisement for Radio Specialties Limited. *The Vancouver Sun*. 1925.

Fig 2.15

than the woman of the house. Instead, radios were staged next to the most beautiful possessions in a wealthy home, such as architectural features, ornate furniture, elaborate chandeliers, musical instruments, drapery, and other finery. The appeal was to class — the wealthy were meant to recognize their worlds in an advertisement that extolled the beauty of the radio cabinet. Long after radio moved out of the cabinet, such cabinets continued to be sold for the purpose of display in the home. In middle class homes, radios were aspirational; their purchase was an extravagance that sometimes became a reality.

Advertising car radio in the post-war period

From its beginnings in the theatre of war, radio was a mobile technology. Early radios were small and frequently powered by batteries rather than electricity, making them easily portable, but as radio was reoriented as a consumer item and models were developed to fit into domestic settings, they became less so.

Advertising for car radios, however, began as early as 1933, even if the desire for one was largely aspirational, realizable by less than ten percent of Canadians.[93] That year, Victor Talking Machine Company of Canada advertised their Victor single-unit auto radio and readers were encouraged to think about "radio as you ride".[94] The advertisement exclaims,

> Motorists! Throw out your clutch and go into neutral for a moment while we give you the biggest news that's been flashed to car owners since four-wheel brakes became standard equipment.[95]

General Electric's 1934 advertisement for its Auto Radio shows a young couple driving along in their convertible, wind running through their hair and the woman's scarf flying along behind them, under the tagline "It's more fun to ride with music" (Fig. 2.15). The copy tells readers,

> Let radio entertainment follow you wherever you travel this summer . . . double the pleasure out of motor trips — picnics and vacations — by having a General Electric Auto Radio installed in your car.[96]

Further down, the advertisement shows three models with comparative features listed, but the contentment on the faces of the young couple is much more important than the technical information once essential to radio advertising. In this advertisement, Canadians are invited to imagine radio in new places and used for new purposes.

The development of car radios took time: first for use of the car to become more widespread, and then for the car to become a standard location for a radio. According to Eric Rothenbuhler and Tom McCourt, "between 1947 and 1962 the radio industry

. . . reconceptualized . . . its audience. Car radios had been available since 1930s; by 1953 nearly 60% of all automobiles were equipped with radios."[97]

In this same period, television was introduced to the home, so radio's role and place shifted to accommodate changing program choices and modes of listening. A negotiation between the media in the home. Radio proliferated — it was brought by the housewife into the kitchen, installed by the husband in the car, and (with the aid of transistors) carried by teenagers onto the streets and into leisure areas.[98] Jennifer Hyland Wang's research on advertising agencies shows that radio maintained its dominance in the home:

> a 1951 advertisement for the CBS radio network read: 'the big advertisers know better than anybody that you don't send a boy to do a man's work. When there's a big job to be done, you'll want radio'.[99]

Nonetheless, radio's dominance gradually eroded as it released network programming to television. It found new niches with new radio models and the introduction of FM broadcasting.

Before the late 1940s, small radios were made portable through the addition of carrying straps. An example is the Canadian General Electric Portable Radio, advertised in *Chatelaine* in 1948 (Plate 2.5).[100] Encouraging the reader to "Carry your Music Wherever You Go", this advertisement depicts a couple in a park on a picnic blanket, food nearby and their portable radio poised between them. Consumers were urged to

> Have more fun on picnics this summer. Carry your entertainment along with you — wherever you go. The new styled GE portables are compact, lightweight radios with a host of new developments that together add up to new peaks of radio performance, new ease and convenience and definitely more value your money.[101]

With the development of the transistor radio in the 1950s, the portable radio took off and designed for fun and practicality. In 1956, *Maclean's Magazine* featured an advertisement that urged readers to "Take your 'Holiday' anywhere…anytime! (Plate 2.6)"[102] The advertisement's text reads:

> It's always playtime with the . . . Westinghouse 'Holiday' 3-way portable radio. Gadabout or play-at-home the Westinghouse 'Holiday' portable spreads fun and music anywhere. Clever new cabinet with sparking two tone styling is as bright and gay as a melody.[103]

In this advertisement, the radio is surrounded by images showing where you could bring your 'Holiday': fishing in a boat, under a beach umbrella, at a picnic, or just lounging at home. The playful design and placement of the text expresses fun, as the transistor radio became associated with youth culture and rising discretionary income of the post-war era.

While mobility was a crucial element of the post-war advertising and development

Fig 2.15 Advertisement for Canadian General Electric Company Limited. *Montreal Gazette*. 1934.

of radio, the symbols of war and the traditions of the home were reinforced in the advertising of radios for the home. During the Second World War, air supremacy, and the military became important symbols prevalent in daily life. Listening to news of the troops and the war became a daily routine in many households around the world, so it is not surprising that a member of the air force is featured in the background of an advertisement for the RCA Victor Globe Trotter Radio. The use of hyperbole to play on the idea of military and radio air supremacy is consistent with the sense of the globe as heard through the radio.

> New RCA Victor Globe Trotter Radio Makes short wave tuning as easy as local! Now you can command the airways of the world. Take flight! Reach new exciting heights of enjoyment... You'll marvel when the new RCA Victor bold Speed overseas Dial tunes in your short wave programs as easily and quickly as . . . local stations.[104]

The ease with which the short-wave stations were reached was essential for families worried about soldiers, volunteers, and family overseas. The advertisement prominently featured the large table model with the updated tuning, but less expensive table top models were also offered at the bottom of the advertisement. The military-like control of the airwaves in the home was offered with the large floor model. The Victor Globe Trotter Radio maintained a presence in the home that was traditional in prior decades when the technology and cultural sensibilities did not allow for the compact design of radio.

Radio in the postwar home

In the immediate post-war period and the decade that followed, radio advertising turned full circle, coming back to its place in a luxurious, well-appointed home. In this spirit, the lady of the household returned in a floor-length gown and long white gloves to announce the natural tone of Northern Electric's "Oriole" in 1946 (Plate 2.7).[105] The cabinet that housed the combination record player and radio is a predecessor of the stereo credenzas of the 1960s and 1970s. The "Oriole" featured a

> cabinet on pleasing modern base . . . of rich American Walnut. . . . Simplified record player and charger is clearly concealed behind the large front panel. The tone of the both the radio and phonograph is natural beyond belief.

The 'natural' tone of this model was reinforced by an image of the oriole and a vase of flowers on the cabinet. Large floor models became more accessible to middle class families in this period, as suburbanization provided access to larger homes and job security made buying on the installment plan possible.

In 1953, Northern Electric's advertising featured its new radio phonograph combination that was "lovely to LOOK at . . . delightful to HEAR! . . . available in walnut, mahogany or limed oak to match the décor of your home." (Plate 2.8)   In the postwar period, the radio/phonograph cabinet was a natural extension of the sleeker modern décor becoming fashionable with young families. Its ability to blend in and match was its most valued attribute.

Philips's "Line of Distinction" marks the end of an era (Plate 2.9). As television arrived in Canada, it was being vigorously advertised, often using the same techniques that had been employed for selling radio. In this advertisement, three women in long gowns surround the Series "400" by Philips. The text reads,

> Heads turn … Admiring eyes light up … for here is the world's most dramatically exciting television — the inspired Series '400' by Philips. Distinguished for its superb character and quality, Series 400 unveils pictures and sound of unsurpassed clarity and brilliance.[106]

Radio would never again dominate the household as it once had.

Advertising imagined and created a place for radio in every Canadian home. Between 1922 and 1960, Canadians fell in love with radio, actively fueled by the advertisers' creation of a role and place for this new medium in the home. As an appliance, its sales far surpassed all other household appliances. Canadians became experts at maintaining their radios and were aficionados of every make and design. As a major purchase, the decision to buy a radio was a weighty one. Radio advertising — largely by radio manufacturers — skillfully created a demand for the technology, as they crafted an image of a world connected by radio. At first a luxury item beyond the means of most Canadians, radio's proliferation into a wide range of models, from table tops to car radios and transistors, made radio ubiquitous in Canadian homes. Technological innovation, scientific mastery, high culture, craftsmanship, beauty, and the possibilities of the world beyond remained mainstays of radio advertising.

The advertisement of radio charted a circuitous route. At first it convinced the experts and hobbyists that preassembled radios could outperform self-assembled ones. As programming expanded, so did the target audience of advertisers. The pleasure of radio programming then became a major selling point, expanding the lives of listeners in new ways with sports, news, music, children's programming, entertainment, and religious broadcasts. To ensure its adoption by the widest number of homes, advertisers positioned radio as a thing of beauty and an essential adjunct to the best homes. Finally, the booming post-war economy permitted radio to expand again, becoming associated with youth culture, cars, and mobility, gradually ceding its dominant spot in the home to television, when it was forced to negotiate its place in the home and lives of its listeners once again.

Plate 2.1 Front cover of Eaton's *Radio Catalogue*. 1927-1928.

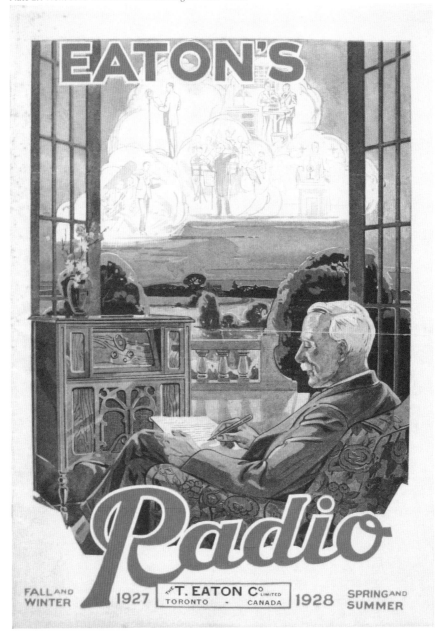

Plate 2.2 Front cover of Eaton's *Radio Catalogue*. 1926-1927.

Plate 2.3 Cover of Eaton's *Radio Catalogue*. 1930-1931.

Plate 2.4 Westinghouse Radio "World Cruiser" advertisement. *Chatelaine*. 1937.

Plate 2.5 Advertisement for Canadian General Electric Co. Limited. *Chatelaine*. 1948.

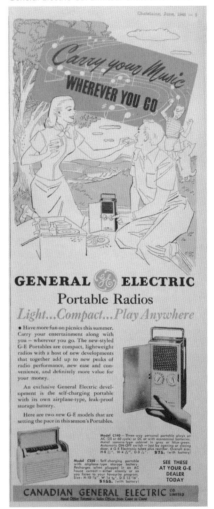

Plate 2.6 Westinghouse advertisement. *Chatelaine*. 1948.

Plate 2.7 Northern Electric Company Limited advertisement. *Chatelaine*. 1946.

Plate 2.8 Northern Electric Company Limited advertisement. *Canadian Homes and Gardens*. 1953.

lovely to LOOK at...
delightful to HEAR!

This new Northern Electric radio phonograph combination is available in walnut, mahogany or limed oak to match the decor of your home. Not only is it lovely to look at, it's a delight to hear — whether you're playing the radio or records. Plays 78, 45 or 33 rpm, records automatically and gives perfect reproduction through the high fidelity wide range speaker. See and hear it at your Northern Electric dealer. Model illustrated — "London" is priced at only $249.50.

YOUR NORTHERN ELECTRIC DEALER IS A GOOD MAN TO KNOW

Northern Electric
COMPANY LIMITED
THE NAME TO REMEMBER FOR ALL HOME APPLIANCES LARGE & SMALL

Plate 2.9 Philips advertisement. *Canadian Homes and Gardens*. 1955.

Notes

[1] Stuart Ewen, "Advertising and the Development of Consumer Society," in *Cultural Politics in Contemporary America*, ed. Ian Angus and Sut Jhally (New York: Routledge, Chapman and Hall Inc., 1989), 82.

[2] Anne F MacLennan, "Learning to Listen: Becoming a Canadian Radio Audience in the 1930s," *Journal of Radio & Audio Media*. 20 (2013): 317.

[3] Canada. Dominion Bureau of Statistics. General Statistics Branch. *The Canada Year Book 1925*. (Ottawa: King's Printer, 1926), 660.

[4] Canada. Dominion Bureau of Statistics. General Statistics Branch. *The Canada Year Book 1932* (Ottawa: King's Printer, 1923), 613; Canada. Dominion Bureau of Statistics. General Statistics Branch. *The Canada Year Book 1934-35* (Ottawa: King's Printer, 1935), 787; Canada. Dominion Bureau of Statistics. General Statistics Branch. *The Canada Year Book 1940* (Ottawa: King's Printer, 1941), xx-xxi, 72 1-722; Canada. Dominion Bureau of Statistics. Information Services Division. *The Canada Year Book 1954* (Ottawa: King's Printer, 1955), 880; Anne F. MacLennan, "Circumstances Beyond Our Control: Canadian Radio Program Schedule Evolution during the 1930s" (PhD diss., Concordia University, 2001), 44-47.

[5] MacLennan 2001, 46; Canada. Dominion Bureau of Statistics. General Statistics Branch. *The Canada Year Book 1940* (Ottawa: King's Printer, 1941), 721-722.

[6] Colin Campbell, "Consuming Goods and the Good of Consuming," in *Consumer Society in American History: A Reader*, ed. Lawrence B. Glickman (New York: Cornell University Press, 1999), 25.

[7] Roland Marchand, *Advertising the American Dream: Making Way for Modernity, 1920-1940* (Berkeley: University of California Press, 1985), 167.

[8] Marconi Long Range Receiver, The Marconi Wireless Telegraph Company of Canada Limited. December 17, 1922. "Radio!" advertisement. *The Vancouver Sun*, 21.

[9] Pierre Bourdieu, Distinction: *A Social Critique of the Judgement of Taste* (Cambridge: Harvard University Press, 1984), 6.

[10] Colin Campbell, 26.

[11] Pierre C Pagé, "L'origine des stations XWA (1915) et CFCF (1922) de Marconi Wireless Telegraph: des données historiographiques à corriger," *Fréquence/Frequency* 5-6 (1996): 151-168; Donald G.Godfrey, "Canadian Marconi: CFCF the Forgotten First," *Canadian Journal of Communication* 8, (September 1982): 56-71; Gilles Proulx, *La Radio d'hier et d'aujourdhui* (Montreal: Editions Libre Expression, 1986), 27; Gilles Proulx, *L'Aventure de la radio au Québec* (Montreal: Editions La Presse, 1979), 22; Marc Raboy, *Missed Opportunities: The Story of Canada's Broadcasting Policy* (Montreal & Kingston: McGill-Queen's University Press, 1990), 5.

[12] Mary Vipond, *Listening In: The First Decade of Canadian Broadcasting 1922-1932* (Montreal & Kingston: McGill-Queen's University Press, 1992) 34; Pagé 160; MacLennan 2001.

[13] Girls and women were also active hobbyists. Michele Hilmes, *Radio Voices: American Broadcasting, 1922–1952* (Minneapolis: University of Minnesota Press, 1997), 132–136; Susan Smulyan, "Radio Advertising to Women in Twenties America: A Latchkey to Every Home," *Historical Journal of Film, Radio & Television* 13 (1993).

[14] Editors of *Electrical Merchandising, How to Retail Radio* (New York: McGraw-Hill Book Company, 1922), 125.

[15] Northern Electric Company Limited. October 14, 1924. "Tune in with the world of Entertainment!" advertisement. *MacLean's Magazine*, 70.

[16] Radiola Super Heterodyne, Canadian General Electric Co., Limited. November 8, 1924. "Radiola Super Heterodyne Receives Stations Thousands of Miles Away Without Aerials or Wires" advertisement. *The Montreal Gazette*, 8.

[17] Radiola Super Heterodyne, 8.

[18] Radiola Super Heterodyne, 8.

[19] Radiola Super Heterodyne, 8.

[20] Radiola III a four tube set, Canadian General Electric Co., Limited. November 15, 1924. "For Long Distance Reception." advertisement. *MacLean's Magazine*, 62.

[21] Radiola III a four tube set, 62.

[22] MacLennan 2001, 12.

[23] Lizabeth Cohen, "Encountering Mass Culture at the Grassroots," in *Consumer Society in American History: A Reader*, ed. Lawrence B. Glickman (New York: Cornell University Press, 1999), 155.

[24] Reynold Wik, "The Radio in Rural America during the 1920s," *Agricultural History* 55 (1981): 340.

[25] Stromberg-Carlson. October 28, 1927. "Making Radio Simple and Sure" advertisement. *Montreal Gazette*, 10.

[26] De Forest & Crosley. October 2, 1925. "The New Radio" advertisement. *The Montreal Gazette*, 8b.

[27] De Forest & Crosley, 8b.

[28] Northern Electric. November 15, 1924. "The Crowning Achievement of a Scientific Century" advertisement. *MacLean's Magazine*, 64.

[29] Artemis Yagou. "Shaping Technology for Everyday Use: The Case of Radio Set Design," *The Design Journal* (2002): 5.

[30] Adrian Forty, in *Objects of Desire: Design and Society since 1750* (London: Thames and Hudson Limited, 1986), 201.

[31] Claude Lévi-Strauss, *The Savage Mind*, 19 in Judith Williamson, *Decoding Advertisements: Ideology and Meaning in Advertising* (London: Marion Boyars, 1978), 110.

[32] Northern Electric. November 15, 1924. "The Crowning Achievement of a Scientific Century" advertisement. *MacLean's Magazine*, 64.

[33] Northern Electric, 64.

[34] Williamson, 116-117.

[35] Benedict Anderson, *Imagined Communities: Reflections on the Origin and Spread of Nationalism*. (London: Verso, 2006).

[36] Atwater Kent Radio. April 1, 1925, "Think What is Back of It" advertisement. *MacLean's Magazine*, 73.

[37] Roland Marchand, 91.

[38] Editors of *Electrical Merchandising, How to Retail Radio*, 122-123.

[39] Stewart-Warner Matched-Unit Radio, Stewart-Warner. October 15, 1925. "The Radio You Have Been Waiting For" advertisement. *MacLean's Magazine*, 35.

[40] Stewart-Warner Matched-Unit Radio, 35.

[41] Stewart-Warner Matched-Unit Radio, 35.

[42] Susan Smulyan, *Selling Radio: The Commercialization of American Broadcasting 1920-1934* (Washington: Smithsonian Institution Press, 1994), 19.

[43] Stewart-Warner Matched-Unit Radio, 35; Smulyan, 19.

[44] General Electric Co. October 15, 1931. "General Electric now presents a 9-Tube Super-Heterodyne" advertisement. *Winnipeg Free Press*, 7.

[45] Northern Electric Company Limited. February 17 1923. "The Children too Can Enjoy Radio" advertisement. *The Montreal Gazette*, 14.

[46] MacLennan 2001, 136.

[47] Noah Arceneaux, "Department Stores and the Origins of American Broadcasting 1910-1931," (PhD diss., University of Georgia, 2007), 145.

[48] Northern Electric Company Limited., 14.

[49] The T. Eaton Company Ltd., Eaton's *Radio Catalogue*. Fall and Winter 1927, Spring and Summer 1928, Front cover.

[50] Richard Butsch, "Crystal Sets and Scarf-Pin Radios: Gender, Technology and the Construction of American Radio Listening in the 1920s," *Media, Culture & Society* 20, (1998): 561.

[51] The T. Eaton Company Ltd., Eaton's *Radio Catalogue*. Fall and Winter Radio Catalogue, 1926-1927, Front cover.

[52] Radiola, Canadian General Electric Co. September 15, 1926. "Tone: The Product of Radio's Master Minds." *MacLean's Magazine*, 31.

[53] Radiola, Canadian General Electric Co., 31.

[54] Radiola, Canadian General Electric Co., 31.

[55] Sparton Radio, The Canadian Fairbanks Morse Company Limited. October 30, 1929. "Face-To-Face Realism: Big Game Reception that is Truly Alive…" advertisement. *The Montreal Gazette*, 14.

[56] Ray Barfield, *Listening to Radio, 1920-1950* (Westport, Conn: Praeger 1996), 11-12.

[57] Anne F MacLennan, "Learning to Listen: Becoming a Canadian Radio Audience in the 1930s," *Journal of Radio & Audio Media*. 20 (2013): 318-319.

[58] Sparton Radio, 14.

[59] Sparton Radio. The Canadian Fairbanks Morse Company Limited. October 16, 1929. "Face-to-Face Realism: Radio Music That Exalts and Thrills…" advertisement. *The Montreal Gazette*, 16.

[60] MacLennan, 2001.

[61] MacLennan, 2001, 17.

[62] Sparton Radio, 16.

[63] Margaret Prang, "The Origins of Public Broadcasting in Canada," *Canadian Historical Review* 46 (March 1965): 4-5; James W. Opp, "The New Age of Evangelism: Fundamentalism and Radio on the Canadian Prairies, 1925-1945," *Historical Papers 1994: Canadian Society of Church History*, 103-104; Mark G. McGowan, "Air Wars: Radio Regulation, Sectarianism and Religious Broadcasting in Canada, 1922-1938," *Historical Papers* 2008: Canadian Society of Church History, 8-9; Russell Johnston, "The Early Trials of Protestant Radio, 1922-38," *Canadian Historical Review* 75:3 (1994): 379; Vipond, 197-198.

[64] RCA Victor Radio. Layton Bros. Limited. August 14, 1936. "Election Day Monday- August 17th" advertisement. *The Montreal Gazette*, 2.

[65] Len Kuffert, *Canada Before Television: Radio, Taste, and the Struggle for Cultural Democracy* (Montreal-Kingston: McGill Queen's University Press, 2016).

[66] "The wireless was even believed by some to be a threat to the privacy of the family, as can be seen in this *Warrington Guardian* editorial (31 March, 1923) entitled 'The Weird Wireless'" as noted in Shaun Moores, "The Box on the Dresser: Memories of Early Radio and Everyday Life, *Media, Culture & Society* 10 (1988): 26.

[67] Jean Baudrillard, *The Consumer Society: Myths and Structures* (Thousand Oaks, California: Sage, 1998), 81-82.

[68] William Boddy, "Archaeologies of Electronic Vision and the Gendered Spectator." *Screen* 35 (1994): 112.

[69] Ralph Sayre, "Putting radio in the parlour," *Radio Dealer* July 1923, 17 as cited in Boddy 1994: 112.

[70] Michael Brown and Corley Dennison. "Integrating Radio into the Home, 1923-1929", *Studies in Popular Culture* 20 (April 1998): 7-8.

[71] Mrs. Hugo Gernsback as quoted in *Radio News*, March 1926, 1373 as cited in Brown and Dennison, 8.

[72] Brown and Dennison, 15.

[73] The T. Eaton Company Ltd., Eaton's *Radio Catalogue*. Fall and Spring Radio Catalogue, 1930-1931, front cover.

[74] Raymond Williams, "Advertising: The Magic System" in Joseph Turow and Matthew P. McAllister, eds. T*he Advertising and Consumer Culture Reader* New York: Routledge, 2009, 13-24; Williamson, 138-151.

[75] The T. Eaton Company Ltd., Eaton's *Radio Catalogue*. Fall and Spring Radio Catalogue, 1930-1931, front cover.

[76] Northern Electric Radio Sets, Layton Bros Limited. February 25, 1925. "Always a Marvel now a Giant for Power" advertisement. *The Montreal Gazette*, 8.

[77] Northern Electric Radio Pioneers. November 17, 1932 "Northern Electric Radio Beauties: These NEW Sets will Amaze You!" advertisement. T*he Montreal Gazette*, 2.

[78] Northern Electric Radio Pioneers, 2.

[79] Louis Carlat, "'A Cleanser for the Mind': Marketing Radio Receivers for the American Home" in *His and Hers: Gender, Consumption, and Technology*, eds. Roger Horowitz & Arwen Mohun (Charlottesville: University of Virginia Press, 1998), 121.

[80] World Cruiser, Westinghouse Radio. October 1937. "Craftsmanship: Only Master Craftsmanship Can Infuse Wood and Metal with this Living Beauty" advertisement. *Chatelaine*, 20.

[81] Michael Windover, *Art Deco: A Mode of Mobility* (Quebec: Presses de l'Université du Québec, 2012), 203-258; Kristina Wilson, *Livable Modernism: Interior Decorating and Design during the Great Depression* (New Haven: Yale University Press, 2004).

[82] World Cruiser, Westinghouse Radio, 20.

[83] DeForest Crosley Radio. Layton Bros Limited. December 11, 1926. "The Merriest Christmas You've Ever Known" advertisement. *The Montreal Gazette*, 10.

[84] DeForest Crosley Radio, 10.

[85] DeForest Crosley Radio, 10.

[86] Bourdieu, 5-6.

[87] Stromberg-Carlson. February 28, 1931. "In keeping with the best in your home" advertisement. *The Montreal Gazette*, 4.

[88] Stromberg-Carlson. December 6, 1929. "Truly Works of Art" advertisement. *The Montreal Gazette*, 11a.

[89] Stromberg-Carlson, 11a.

[90] Michael Brown and Corley Dennison. "Integrating Radio into the Home, 1923-1929", *Studies in Popular Culture* 20 (April 1998): 11.

[91] Fada Radio, Radio Specialties Limited. October 14, 1925. "Cabinets that Harmonize with Beautiful Interiors" advertisement. The Vancouver Sun, 10.

[92] Fada Radio, 10

[93] In 1933 when the Victor Auto Radio advertisement appeared, 1,082,957 Canadians were registered automobile owners. "Table 36. Numbers of Motor Vehicles Registered in Canada, by Provinces, calendar years 1907-33," Canada. Dominion Bureau of Statistics. Department of Trade and Commerce. *Canada Year Book 1934-35*. (Ottawa: King's Printer, 1935), 738. The total population of Canada in 1931 was 10,376,786; Table 1. "Population of Canada, by Provinces and Territories, in the Census years 1871-1931," Canada. Dominion Bureau of Statistics. Department of Trade and Commerce. *Canada Year Book 1934-35*. (Ottawa: King's Printer, 1935), 99.

[94] Victor Single-Unit Auto Radio. Victor talking Machine Company of Canada, Limited. June 27, 1933. "Victor Single-Unit Auto Radio: Victor Single-Unit Auto Radio" advertisement. *The Montreal Gazette*, 6.

[95] Victor Single-Unit Auto Radio, 6.

[96] Auto Radio, Canadian General Electric Company Limited. June 28, 1934. "It's More Fun to Ride with Music" advertisement. *The Montreal Gazette*, 10.

[97] Sterling Table 670-A as cited in Eric Rothenbuhler and Tom McCourt, "Radio Redefines Itself, 1947-1962," in *Radio Reader: Essays in the Cultural History of Radio*, eds. by Michele Hilmes and Jason Loviglio, (New York: Routledge, 2002), 378.

[98] McCann-Erikson, "The Case of the Radio-Active Housewife" as cited in Jennifer Hyland Wang. "The Case of the Radio-Active Housewife," in *Radio Reader : Essays in the Cultural History of Radio,*. eds. by Michele Hilmes and Jason Loviglio, (New York: Routledge, 2002), 361.

[99] "Network Radio." *Sponsor* July 16, 1951: 46-47 as cited in Wang, 361.

[100] General Electric Portable Radios. Canadian General Electric Co., Limited. June 1948. "Carry Your Music Wherever You Go" advertisement. *Chatelaine*, 5.

[101] General Electric Portable Radios, 5.

[102] "Holiday" 3-way portable radio, Westinghouse. May 26, 1956. "Take your "Holiday" anywhere…anytime!" advertisement. *Maclean's Magazine*, 42.

[103] "Holiday", 42.

[104] RCA Victor Globe Trotter Radio, RCA Victor Company Limited. November 1940. "Symbol of Air Supremacy New RCA Victor Globe Trotter Radio Makes Short Wave Tuning as Easy as Local" advertisement. *Chatelaine*, 59.

[105] Northern Electric Oriole, Northern Electric Company Limited. December 1946. "Northern's Natural Tone" advertisement. *Chatelaine*, 27

[106] Philips. December 1955. "Line of Distinction" advertisement. *Canadian Homes and Gardens*, front cover inside liner.

Electrohome pamphlet. 1947.

RHAPSODY

This striking design by Deilcraft offers a choice of either an 8-tube FM/AM or a 6-tube dual wave chassis. Other Rhapsody features are—large 12" permanent magnet speaker—continuously variable tone control—bin type record changer compartment—built-in record space—built-in and outdoor FM antennas double as AM aerials—cabinet by Deilcraft. Walnut and Autumn Leaf Mahogany. 31¾" H, 32" W, 17⅝" D.

Situating Radio:
How Radio Changed Canadian Space

3

Radio created a new kind of social space for Canadians.[1] The electronic medium reinscribed national space through new public-making institutions such as the Canadian Broadcasting Corporation, and affected intimate settings like the home and automobile, enriching them with a spatial quality of being 'there' and 'elsewhere' at the same time.[2] The unique sense of intimacy that radio conveys, which has been explored in both the American and Canadian contexts, carried connotations of proximity despite vast geographic distance.[3] Other tropes in on-air programming and even the selection of material being broadcast from distant locations (e.g., the New York Philharmonic heard in rural Saskatchewan) emphasized that this modern medium compressed time and space.[4] This chapter looks at how radio space was created in a 25-year period beginning in the mid-1920s, examining advertisements, radio consoles, floor plans and images from popular magazines, as well as buildings designed to broadcast or transmit radio. By the end of the Second World War, radio space — formed through the architecture and material culture of the medium — was an indispensable part of Canadian society.

Radio had a rich visual and material culture that helped users visualize its aural qualities and its spatial dimensions. For instance, a 1937 advertisement for the latest Viking models in a radio catalogue for the T. Eaton Company evokes the way the electronic medium was cast as a time-and- space-altering instrument (Fig. 3.1). The taglines "Go Visiting with Viking" and "Make the World your Neighbor," are juxtaposed with handsome, walnut-veneered receivers on either side of two women having a conversation across a globe. The message is clear: radio can connect people across great geographic expanses in a familiar, neighbourly way (and Eaton's, the great Canadian department store that both manufactures and sells these sets, can facilitate this modern mode of communication). Here, radio listening is depicted as an activity. Media historian Kate Lacey argues that listening ought to be conceptualized as a form

Fig 3.1

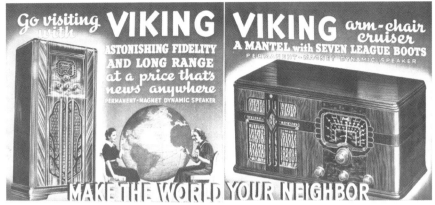

Battery-Operated Long and Short Wave Models   We Know of No Better Value at the Price

of engagement, and is in fact crucial to the creation of publics.[5] Even the name of this line of radios, "Viking", suggests travel, adventure, and the conquest of space. Most importantly for the purposes of this chapter, the advertisement signals a significant change in the concept of social space brought about by radio. Once incorporated into everyday life activities, radio offered a new way to link people together, both intimately (through radio's address) and spatially (within a network of dispersed apparatuses).

A 1929 advertisement marking the opening of the Canadian Pacific Railway's luxurious Royal York Hotel in Toronto offers another useful perspective on the creation of radio space in the country (Fig. 3.2). In the advertisement, the presence of radio reception in every room indicates up-to- the-minute luxury.[6] The gentleman in the image reclines with a cigar in hand as he listens to a broadcast through headphones. Below, the copy describes how visitors can tune into musical performances and other events taking place in the hotel, and that a children's playroom is equipped with a loud speaker to amplify special programming for them. The Royal York, the largest (and perhaps most modern) hotel in the British Empire, is here a site of broadcast and a locus of reception. Images of other hotels, ships, and railroads through Canadian landscapes remind readers that this luxury hotel in Toronto is part of a vast, national, and imperial network of transportation, commerce, and leisure. The radio infrastructure, in this case supplied by Northern Electric, complements and enhances this sense of connectivity.

Canadian Pacific Railway's competitor, Canadian National Railway (CNR), may have spurred on the incorporation of radio service. Sir Henry Thornton, president and chairman of CNR, founded the first nation-wide radio network in North America back in

Fig 3.1 Viking Radio advertisement from Eaton's *Radio Catalogue*. 1937-38.

Fig 3.2 Northern Electric advertisement. *MacLean's Magazine*. 15 May 1929.

Fig 3.3

1923, providing service to the company's rail passengers as well as anyone else within proximity to the transmitters.[7] Passengers could listen to company sponsored or created programs in a coach car or a hotel room (Fig. 3.3). Many of CNR's studios were located in its hotels, including Ottawa's Château Laurier (Fig. 3.4). In creating this radio network, CNR built its broadcasting capacity on the armature of its telegraph system. This meant that an operator in Toronto could coordinate a program from Ottawa to be broadcast simultaneously in Winnipeg.[8] This radio infrastructure was transferred to the Canadian Radio Broadcasting Commission in 1933 and then to CBC, with its formation in 1936.[9] From its outset, radio — as a nationalizing project in Canada — was inextricably bound to transportation infrastructure of cross-continental railways. Both commercial and

Fig 3.4

political networks were thus essential to the production of radio space.

Harold Innis studied the role of railways in the formation of Canada and Canadian culture, and late in his career developed communications theory that emphasized the spatial and temporal effects of media.[10] For Innis, radio was a space-biased medium that assisted monopolies of power to control territory. His junior colleague in the English Department at the University of Toronto, Marshall McLuhan, developed some of Innis's ideas further and likewise recognized the power of media to produce publics, or what political anthropologist Benedict Anderson later called "imagined communities."[11] It is perhaps not surprising that both Innis and McLuhan picked up on the spatio-political effects of media, as Canadians witnessing and weighing in on debates about Canadian

Fig 3.3 Passengers listen in aboard Maple Leaf radio car. 1929.

Fig 3.4 CNR performers in CNRO studio. 1926.

culture in the face of American cultural imperialism and the creation of a national network of electronic communications in a country as vast and thinly populated as theirs.[12]

With their insights in mind, this chapter will examine the form of the medium rather than on-air content.[13] It will consider the design of radio consoles and surrounding visual culture, which suggested ways of positioning and using the apparatus at home. And it will examine spaces as intimate as the automobile and as public as large-scale institutional architecture. The visual and material culture of radio reminds us of how rich and dynamic radio culture was in this period and how spatially entrenched and thus normalized it had become by the mid-1950s.

Placing Radio Communities in the 1920s

The Radiola III was released in 1924 and was manufactured by Canadian Westinghouse in Hamilton, Ontario (Plate 3.1).[14] The compact receiver — a mahogany box, with black synthetic cover and dials, and two glass vacuum tubes — provides an example of moderately priced radios that began to populate Canadian homes in the early-to- mid 1920s.[15] The device has the look of a scientific instrument — uncluttered and precise. Advertisements for it lauded its capacity to receive signals from 1,500 miles away.[16] With only two vacuum tubes to detect and amplify radio signals, this regenerative receiver was certainly more effective than cheap crystal sets, but not powerful enough for amplification through a loudspeaker.[17] Instead, radio audiences would use earphones to "listen in," like the passengers of the CNR radio car.

Radio listening with this device was generally a solitary activity, unless earphones were placed in a porcelain bowls, pots, etc., to amplify sound.[18] Early users often engaged in DXing, tuning in and recording the call numbers of distant stations.[19] Thomas Everrett, in his study of headphones, points out that the very careful listening practice of DXing required headsets.[20] This individualist activity often sidelined women, who were portrayed in some cartoons of the era as "wireless widows".[21] Some models of this era included multiple headphone inputs, providing a communal yet physically separated socio-acoustic experience.[22]

The Westinghouse Radiola III (Fig. 2.6) could be purchased through the T. Eaton Company's radio catalogue, with prices listed for full sets including delivery. Commercial networks like Eaton's assisted in the mapping of radio space. Its distribution channels provided equipment to receive and power radios. Eaton's catalogues are a rich source of information about radio culture, providing insight into how radio was perceived in Canada.[23] The 1924-25 catalogue for example, includes full sets like the

Radiola III, as well as parts and blueprints for do-it- yourselfers, suggesting an audience of both experienced enthusiasts likely engaged in DXing, as well as a more "polite" audience interested in bringing radio entertainment into the living room.

A comparison of Eaton's catalogue covers between 1924 and 1925 indicates a growing middle-class audience for radio (Plates 3.2 and 3.3). In the earlier one, a radio transmission tower suggests that the new medium provides access to many geographical regions, presented in the surrounding vignettes, with the back cover listing North American radio stations that a proficient user might be able to tune into. The emphasis is on radio transmission and its connecting spaces. The 1925 cover, however, is more focused on reception. In a larger format and in full colour, two well-dressed women are listening together through a gooseneck loudspeaker. The radio on display is likely the Neutrodyne Set offered as a blueprint and with a list of required parts that could be shipped and assembled at home. The back cover (Plate 3.4) depicts the interpenetrating spaces of the radio system, including broadcasting studio (reminiscent of the studio in the Château Laurier), transmission towers, a radio operating room, and the home receiving set. This representation helped introduce radio visually and spatially to those unfamiliar with the new technology.

The presence of the speaker is significant. The mid-1920s saw the development of more effective speakers, which helped to domesticate the new technology. Early loudspeakers used telephone receiver technology (commonly employed in headphones) at the base of the horn.[24] By 1927, a new dynamic, moving-coil or cone speaker amplified sound without distortion (Plate 1.2).[25] Together with the advent of alternating current-powered tubes, such as those pioneered by Ted Rogers around 1925, the radio could be placed in the living room or parlour without fear of leaky batteries damaging carpets and furniture in, what was most likely an urban home, given that it was supplied with electric power.[26]

Advertisements for Rogers' batteryless radio that operated from an electric light socket explicitly appealed to a broad consumer base (Fig. 3.5). "Just plug in, then tune in!" goes the tagline, and the copy explains that "With a Rogers radio you do not need to understand Radio science any more than you need to be a mechanic to own a Rolls-Royce automobile." The Rogers Batteryless Model 110 combines an enclosed speaker with A/C powered tubes (Plate 3.5). Radio listening, which had been in many instances solitary or semi-private (e.g., people listening together through multiple sets of earphones) and more associated with masculine spaces such as the garage, attic, or basement, moved into the public spaces of the home.[27] And with this, radio's public grew.

Fig 3.5

Fig 3.5   Rogers advertisement. *MacLean's Magazine.* 1 Nov 1925.

Fig 3.6

With its establishment in the living room, radio began to appear in home decorating magazine ideal floor plans, indicating an acceptance of the electronic instrument in the home (Fig. 3.6). The first time a furnishing plan included a radio in *Canadian Homes and Gardens*, the designer represented a living room for upper-middle- class consumers "anywhere in Canada". The plan is described as "simple" and "logical", with groups of furniture "arranged for ease of conversation, and with ample clear space in the centre of the room." The radio ('H' in the plan) is not the focal point of the room — the traditional fireplace continues to hold this honour — but is positioned next to the window. The Philco superheterodyne highboy chosen is described as "[t]he indispensable contact with the outside world".[28] Visually linked with the outdoors by its placement next to a window, the imaginative and space-compressing potential of the radio is underlined in the scheme.

The radio receivers pictured in *Canadian Homes and Gardens* are substantial pieces of fine furniture.[29] The Stromberg-Carlson Model 25-A of 1932 from the collection of the Canada Science and Technology Museum is an example of this type (Plate 3.6). This floor console was a prestigious object, carefully carved and detailed in a William and Mary style.[30] Its design would have resonated with other more traditional furniture styles prevailing in Canadian living rooms at the time. Compared to the Radiola III's frank display of vacuum tubes, the Stromberg-Carlson console downplays evidence of the electronicmedium, its speaker grille, dial, and knobs blending into the decorative woodwork. Its unique iconography however, points to its auditory function. Garlands of leaves frame the grille in the form of a proscenium arch, and open curtains hang from three intersecting lyrebird feathers.[31] The "v" shape, reminiscent of a lyre, is repeated on the upper corners of the front. The use of lyrebird feathers not only suggests music and aural representation — lyrebirds are known for their ability to mimic sounds from their environment — but perhaps also an imperial connection, since the birds are native to Australia. This may have had resonance at the time, since Canada was still very much aligned with the British Empire, hosting the Imperial Economic Conference in 1932, which, coincidentally was the first year of the BBC's empire service.[32] The Stromberg-Carlson model 25-A, which linked listeners to the wider world, was clearly an object to be seen and admired. This was no intrusion into the family living room, but a well-crafted possession worthy of display in the best and most public room of the house. This model underscores how radio had become an aesthetic object. It was meant to be touched, and its profusion of ornament highlights and encourages a tactile experience. While not all floor consoles of the period had this level of detail, the highboy form was quite popular, indicating a general trend for electronics to be encased in fine furniture.[33]

Radio was conceptualized thus not only as a gateway to the wider world, but as an ornament and an integral part of the modern living room. A 1930 ad campaign for Marconi emphasized the connection between radio and interior design. Appealing to the expertise of Mrs. Minerva Elliot, a leading Canadian interior decorator and regular contributor to *Canadian Homes and Gardens*, the new radio was described as "a decorative asset to any room".[34] This deluxe and expensive model, which included both phonograph and radio, was obviously aimed at an upper-middle-class audience who would recognize Elliot's name. The copy underscores the idea of fashion and interior design, noting that new models would "[add] dignity and lasting beauty to the furnishing scheme" of any room, and, importantly, that consumers would replace their old (current) sets with this one, once they had heard and seen it.

As the 1930s progressed, radio continued to be featured in ideal living room floor plans, rather than the kitchen or bedroom (Fig. 3.7). In a 1934 article in *Canadian Homes and Gardens*, Whitney Dill includes a "radio of finely grained walnut" in a room comprising a blend of Directoire and Biedermeier furnishings with a "Classical-

Fig 3.7

Fig 3.6 Article from *Canadian Homes and Gardens*. June 1931.

Fig 3.7 Plan from *Canadian Homes and Gardens*. Oct-Nov 1934.

Fig 3.8

Modern" sensibility.[35] The radio, located on the upper wall of the living room, appears to be situated in a niche, framing the receiver as a secondary focal point in the room, after the fireplace. Its proximity to the hearth creates a coherent zone by linking the two. The space represented here conjures nostalgic images of families sitting together listening to popular programs.

However, while ideal floor plans of *Canadian Homes and Gardens* may have influenced Canadians of means, for a good many middle-class Canadians, mantel models were the more affordable option during the Depression. The Philco 20 "Baby Grand" of 1930 was a popular early example of this type. Less ornamented than the later Stromberg-Carlson console, the Baby Grand sported Gothic tracery on its grille (Plate 3.7). Not only did this model blend well into Canadian households, its allusion to Gothic architecture may have carried religious overtones for the predominantly Christian population of Canada at the time, especially during religious broadcasts. The name "Baby Grand" of course evokes an association with the piano, an important piece of furniture in many middle-class living rooms.[36] An advertisement in the *Vancouver Sun* emphasized the listener's control over the medium, like a conductor or musician:

> With Philco Tone Control you wield a conductor's baton on every program. At your finger tips — to meet your mood of the moment — to interpret music, song or speech the way you want it — is radio's newest miracle Philco tone Control. Pensive? Turn the Tone Control to mellow or deep. Gay? Turn it to brilliant or bright, and like the thousands of other Canadians who have bought a 1930 Philco you will say 'At last radio has been made as expressive — as personal — as a violin.'[37]

Here, the radio is cast as a democratic instrument, available to everyone as its price drops while offering a level of participation in the medium.[38] But the most important feature of the Baby Grand, besides its price, was its size. At 17-1/2 inches high and 16 inches wide, it was easier to accommodate in the kitchen or bedrooms than were the larger floor models.[39] This created opportunities for multiple listening spaces in the home.[40] With the growth in listening publics and the broad acceptance of the radio in Canadian homes, the stage was set for radio — a modern technology — to help usher in modern design both in domestic and public spaces.

Radio and Modern Design in the 1930s

Radio was unequivocally sold as modern. Even advertisements for other products, such as hardwood flooring, pointed to radio culture as heralding new kinds of contemporary activities, like dancing to popular music (Fig. 3.8). Radio thus had broader spatial implications than stationary listening. Modern design was often associated

Fig 3.8 Seaman-Kent Hardwood Flooring advertisement. *Canadian Homes and Gardens*. April 1928.

Fig 3.9 General Electric advertisement. *Winnipeg Free Press*. 1931.

Fig 3.10 Marconi advertisement. *Montreal Gazette*. 1930.

with movement and speed.[41] Simple lines and abstraction became a shorthand for modernity in radio advertising and cabinet designs. A Canadian General Electric (CGE) advertisement from 1931 provides a good example of how consumers — and their spaces and behaviours — were represented as assuredly modern (Fig. 3.9). A high contrast image of mother, father, and daughter sit in front of a typical highboy console. The couple's fashionable dress suggests they are people in tune with their times. The radio cabinet, left a gleaming white and set against a stepped black silhouette, is flanked by two Art Deco wall sconces. Like the ads for the Baby Grand, the CGE Junior Consoles are marketed as powerful, petit, and affordable. The crisp lines of this advertisement, however, suggest that radio listeners are modern.

While advertisements signalled modernity through typography and graphics, the use of modern design in radio cabinets was more cautiously and slowly adopted.[42] The Canadian Marconi Company linked modern design to radio more indirectly by pointing to its newly opened manufacturing plant in Mont Royal (Fig. 3.10). In the copy, the company claims to have consulted musicians, interior decorators (and this is part of the same campaign that featured Mrs. Minerva Elliot), and scientists, proving its modern design principles. An image of the façade of the new plant commands attention, while text underneath notes that fifteen sets are assembled every half hour inside.[43] The advertisement pointed to another space of radio: the manufacturing plant. The Art Deco façade represents the functional approach and efficiency of the company, which is embodied in its consoles. Even if the receivers themselves lack a modern outward appearance, consumers could associate themselves with the modern, public face of the company.

In the year following this advertisement, companies began offering models in a *moderne* mode in Canada.[44] As architectural and design historian Adrian Forty pointed out in the case of England, radio became a key conduit for bringing modern design to the masses, and indeed was more effective in its popularization than most architectural and design discourse at the time.[45] An early example of this move to Art Deco design in Canada is the Westinghouse Columaire, designed by Raymond Loewy in New York but manufactured for Canadians in Montreal (Plate 3.8).[46] Like the Baby Grand and CGE Junior Console, the Columaire was designed to meet the demand for a powerful radio with a reduced footprint. But unlike these models, the Columaire attempts to give radio a new form. Occupying one square foot of floor space, its design merges a traditional form (the grandfather clock) with the silhouette of contemporary skyscrapers. And according to the advertising copy, the design follows the "column of air principle" (with the speaker oriented upward and the top covered with grille cloth).[47] It describes

Fig 3.9

Fig 3.10

Fig 3.11

—Drawn for *MacLean's* by W. Gordon Wallace.

"*Pardon me, have you the time?*"
Radio Enthusiast: "*When the gong sounds it will be exactly six fifteen.*"

the model as "up-to- the- minute- modern as the ninety story skyscraper", cleverly linking both the formal appearance to the secondary function (electric clock). Indeed, it appears that Loewy, like other designers including Paul T. Frankl, was inspired by the skyscraper and aimed to represent its form in furniture designed largely for apartment living (perhaps even in one of the new high-rises).[48] Designed for modern lifestyle, the Columaire was given a form that represented and responded to the spatial implications of contemporary urban life.

The model also underlined radio's connection to time. With the rise in popularity of the new medium and the advent of national networks (especially in the USA), the amount of content on air grew and became more specialized (e.g., children's and women's programming in the daytime).[49] A cartoon by W. Gordon Wallace in *Maclean's Magazine* draws attention to radio's regimented time (Fig. 3.11). Radio reinforces standardized time and is thus associated with precision, not to mention modernity — with fashionable clothes and the front end of an automobile visually buttressing this notion.

In some ways, this cartoon, and indeed the Columaire, underlines how radio as "representational space" reinforced a "representation of space," to use philosopher Henri Lefebvre's terms.[50] Social space, measured by standard time, structures spatial practices, such as going to work, seeing a movie, or listening to a scheduled radio program. The Columaire takes this even further by collapsing the clock (keeper of time) with the radio (maker of the activity of radio listening) in a form that represents stable,

Fig 3.11 Cartoon in *Maclean's Magazine*. 1931.

Fig 3.12 Westinghouse Columaire advertisement, detail. *Winnipeg Free Press*. 1931.

Fig 3.13 Article in *Maclean's Magazine*. 1936.

Fig 3.14 Canadian General Electric advertisement. *Canadian Homes and Gardens*. 1936.

modern society (silhouette of a skyscraper).⁵¹ Ultimately, an underlying system of organization — a "representation of space", in Lefebvre's terms — finds representational form and produces spatial practices.

Another point of intersection between radio and spatial practice is the way the medium could be used to regiment the body. For instance, the Columaire's sleek profile was emphasized in some advertisements and was compared visually to the slender body of the modern woman (Fig. 3.12). Like the reduction of detail in favour of abstracted forms in the CGE Junior Console ad, the design of the Columaire follows the modernist principle of formal simplicity. An article from *Maclean's Magazine* five years later shows how radio could be used to discipline the body in pursuit of a fashionable silhouette (Fig. 3.13). Scholars such as Christina Cogdell have explored the links between research into ideal bodies (eugenics) and streamlining, which were both popular in 1930s North America.⁵² Form, fashion, and fitness of purpose all converge in this facet of radio culture.

By the mid-1930s, Art Deco radio sets, whether smaller mantel or more expensive floor models, appeared regularly in catalogues and advertisements. By this time radio had become well established in the living room and signified modernity. For instance, in an advertisement for Dominion Inlaid Linoleum, the console and *moderne* furniture depicted suggest that the space represented is a living room, a room that previously may not have been a candidate for this type flooring (Plate 3.9). With a subtle reference to the radio, the copy asks consumers to "Let the colour magic of Dominion Inlaid Linoleum put your home in tune with the times. Linoleum will reduce housework and give you greater leisure." The flooring product is as modern as the radio, and its installation in a *moderne* living room like this allowed home owners more leisure and listening time. The presence of a receiving set in an advertisement for linoleum is significant for its evocation of spatial practices and its representation of radio as a marker of social space. Here, radio anchors the furnishing scheme and lifestyle.

Not unsurprisingly given its context, a radio also anchors a contemporaneous advertisement for the "General Electric Living Room" (Fig. 3.14). Positioned below a modern landscape painting (a kind of window onto the world, in keeping with the established convention of placing radios near windows to evoke connection to other spaces) and with a sculpture of a bird about to take flight on its top, the radio once again represents modernity and a gateway to the wider world. The CGE model here stands for modern leisure and is equated with, or even elevated above, the status of the piano located behind the women to the right. With the all-electric house like the one featured here decades away from a middle-class reality, this almost Hollywood set-like

Fig 3.12

Fig 3.13

Fig 3.14

Fig 3.15

image is clearly aspirational.[53] The well-dressed folks strike poses not dissimilar to CGE advertisements from five years earlier, but now the entire stage is dressed modern (*or moderne*), and the radio joins soft lighting and air conditioning to produce a healthy, relaxing, and truly modern space.

Turning our attention from the home to the design of the radio station, we see a much greater willingness to embrace modern design.[54] A prime example is the transmission station built by the CBC in Hornby, Ontario (Fig. 3.15).[55] Shortly after its establishment in 1936, the CBC began building regional high-power stations to develop its network, beginning with nearly identical buildings in Verchères (outside Montreal) and Hornby (outside Toronto). As noted above, the CBC had inherited properties of the CRBC (and earlier CNR's radio facilities), which included some purpose-built facilities, but for the most part radio studios were tucked into older buildings like the Château Laurier in Ottawa. CBC's new transmission stations gave its chief architect David Gordon McKinstry and his team the opportunity to build something which represented the new medium and

Fig 3.16

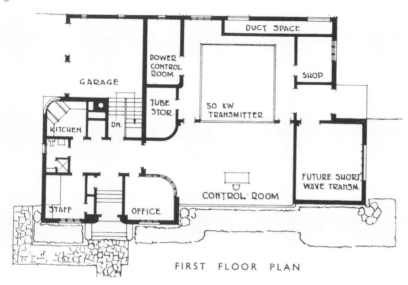

the crown corporation. As Mackenzie Waters notes in the *Journal, the Royal Architectural Institute of Canada (JRAIC)*, radio stations offered opportunities to build in a modern mode, since the typology was unfettered by tradition and the electronic medium was clearly associated with modernity.[56] The decorative "speed lines," abstracted CBC logo, and glass bricks visible in Figure 3.15 enlivened the surface of the otherwise fairly simple, concrete box. The streamlined building would have resonated with other popular modern locations on main streets in the US and Canada, such as cinemas and post offices, but it also stood for a new public institution.[57] Streamlining, which represented a fascination with the modern space of fast cars and airplanes — in short, progress —appeared quite appropriate for this new institution. The CBC Hornby station was a unified structure with the decorative elements from the exterior repeated in its interior.[58]

McKinstry and his architecture and engineering colleagues designed the station to dramatize radio operations. As visitors entered, they ascended a few steps to turn right, where they could see the control room from behind a curved railing (Fig. 3.16-3.18). On the floor in front of them, a map of Canada marked the radio stations across the country.[59] Here, the CBC is advertising itself as a material network of stations, much like railway stations of the CPR or CNR. Radio assumes a similar task of nation building, collapsing time and space by bringing publics from across the country together in one acoustic space.

Fig 3.15 CBC transmission station (CBL). Hornby, ON. 1937.

Fig 3.16 CBC transmission station (CBL). Hornby, ON. 1937. First floor plan.

Fig 3.17 CBC transmission station (CBL). Hornby, ON. 1937. Interior view of control room.

Fig 3.18 CBC transmission station (CBL). Hornby, ON. 1937. Interior view of operator in control room.

Radio Heterotopias

The construction of high-power transmission stations reinforced national boundaries, since frequencies were allocated to nations through international agreements and were conceived of by some as national resources. The transmission stations also provided consistent access to on-air content over a large region.[60] Two of radio's unique spatial qualities are being able to tune in, no matter where you live; and being able to "travel" nearly anywhere while remaining at home. Advertisers and manufacturers made the most of these attributes in selling radio.[61]

The very name of some models, like Westinghouse's "World Cruiser" or RCA Victor's "Globe Trotter", made apparent the link between travelling and radio. The RCA Victor "Globe Trotter" Model C6-1 in the Canada Science and Technology Museum's collection (Plate 3.10) blends notions of stability and firmness appropriate to its place in the living room (evinced by the fine veneered Art Deco cabinet) with a space-conquering instrument, suggested by its large round dials and image of the globe in the tuning window. In some ways, the design of this cabinet evokes the radio listeners' experience of being both at home and elsewhere at the same time.

The phenomenon of inhabiting or juxtaposing different spaces at the same time resonates with philosopher Michel Foucault's notion of "heterotopias."[62] In "Of Other Spaces," a lecture given to architects in 1967 and posthumously published, Foucault describes heterotopias as real places (unlike utopias) that

> have the curious property of being in relation with all the other sites, but in such a way as to suspect, neutralize, or invert the set of relations that they happen to designate, mirror, or reflect.[63]

According to Foucault, a heterotopia can juxtapose several, sometimes incompatible spaces in a single place. While he refers to cinema as an art form that accomplishes this, radio might also be understood along these lines. When sounds from different places enter the space of the home, a heterogeneous mixing of spaces occurs. Foucault was interested in such "other spaces," seeing them as key sites for the management of crises and places imbued with political valence, even, in some cases, spatial violence.[64] We might see the RCA Victor "Globe Trotter" (and any radio from this period) as mediators and societal pressure valves, simultaneously providing access to the world while circumscribing it, managing the potential of social upheaval through content and forms that were socially conservative.[65] The 1936 Layton Brothers advertisement discussed in chapter two, which was selling RCA Victor radios with an appeal to the Quebec election, demonstrates quite directly how the radio console was the listeners' conduit to a political forum (Fig. 2.11).

Fig 3.19 Northern Electric advertisement. *Montreal Gazette*. 1936.

Fig 3.20 *Philco Radio Atlas of the World*, 1936.

Foucault closes his lecture "Of Other Spaces" by arguing that a boat is a heterotopia *par excellence*:

> a floating piece of space, a place without a place, that exists by itself, that is closed in on itself and at the same time is given over to the infinity of the sea. . . . 'the greatest reserve of the imagination'.[66]

The medium of radio was certainly associated with imagination,[67] and consoles were sometimes compared to ocean liners, including Northern Electric's series from 1936 (Fig. 3.19). The "Normandie" refers to the French ocean liner which entered service in 1935 and was the largest and fastest passenger ship making the crossing from Europe to New York. The name evoked technological superiority, speed, elegance and, of course, the power to travel the world.

Fig 3.19

The idea of "travelling" by radio conferred on the medium a cosmopolitan character, especially as shortwave receivers became more common in the mid-1930s. The cover of the Eaton's *Radio Catalogue* from 1935-36 shows this effectively (Plate 3.11). In the middle, a dial with a globe in the centre (not dissimilar to the dial of the "Globe Trotter") depicts standard broadcasting frequencies and international shortwave bands. Surrounding it are people from around the world, suggesting places and cultures that a listener— represented by the hand reaching in from below—could access with the turn of a knob. This radio catalogue, like the receivers it is advertising, offers "The WORLD at your fingertips".

With the development of a shortwave receiver market, manufacturers produced radio maps and atlases, which helped listeners visualize "radio space" (Plate 3.12 and Fig. 3.20). Radio's ability to make electronically mediated encounters instantaneous likely inspired McLuhan to conceptualize the effects of media in spatial terms. By situating people in specific locations and contexts while simultaneously linking them with a global connection to other places, radio created a unique modern space, or what McLuhan would later call the "global village."[68]

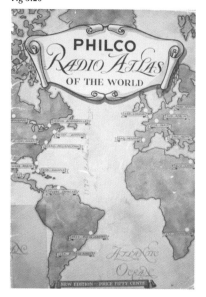

Fig 3.20

Pre-set tuning buttons on models such as RCA Victor A-1 of 1939 also helped to situate listeners in space (Plate 3.13).[69] The development of push-button controls indicated the growth in numbers of powerful radio stations as well as how radio models could reinforce one's geographical position. For instance, the RCA Victor A-1 has buttons set for CFNB (Fredericton, NB), WOR and WABC (New York), WBZ (Boston), CBA (Sackville, NB), and CHSJ (Saint John, NB). The listener most likely lived in New Brunswick and thus had access to local stations in Fredericton and Saint John, not to mention Sackville, which had opened in 1939 as the CBC's powerful Maritime regional station. But the listener could also drop into Boston or New York easily to listen to the

Fig 3.21

Fig 3.22

Mutual Broadcasting System (with WOR), NBC (eventually Red Network with WBZ), and ABC (originally Blue Network of NBC with WABC). While the pre-set buttons both located this radio in space and oriented it toward a pattern of radio listening (i.e., listening frequently to these particular channels and perhaps jumping between them on occasion), the shortwave and police bands provided other possibilities. This model is also interesting for its quarter-inch input in the back, designed for incorporation with a phonograph or RCA's new televisions, which were being introduced at the New York World's Fair in 1939-40.[70]

While radios with pre-set buttons signalled patterns of regionally specific radio listening, the advent of the automobile radio in the late 1920s and its growth in the Canadian market in the 1930s created perhaps the ultimate heterotopic experience for radio, somewhat akin to Foucault's boat.[71] In an era before car radios were standard, some advertisers encouraged radio enthusiasts to bring portable receivers on the road.[72] Auto radios, comprising speaker with receiver and tuner and antenna outside, however, increasingly appeared in daily newspaper advertising (Figs. 3.21, 3.22). According to Butsch, by 1941 nearly a third of all cars in the USA had auto radios and fifty percent of new automobiles were equipped them.[73] As historian of technology

Karin Bijsterveld points out, early auto radios worked best when the car was not moving.[74] With improvements in design, including incorporation of the radio into the car's battery, radios became more common, and advertisements for RCA and Philco, for instance, depict men and women driving while listening. Cars became "mobile listening booths".[75] For instance, the CBC carried out a field reception survey for its Prairie regional high-power transmission station in Watrous, Saskatchewan by automobile.[76] The high-power stations erected by the CBC helped automobile listeners stay connected, affording that heterotopic experience of travelling through the Canadian landscape while virtually inhabiting other spaces.

Radio during the War

The CBC's high-power stations in the late 1930s helped unify the country through a consistency of national radio space.[77] The CBC's own architecture department designed all of the stations, providing visual similarities across the country. The locations of the stations were proudly displayed at the reception desk in a CBC operated building in Montreal (Fig. 3.23).[78] Shortly after the erection of the first four high-power stations in Saskatchewan, Ontario, Quebec, and New Brunswick, the Second World War erupted, making radio even more strategically significant. Indeed, it was the CBC's coverage of

Fig 3.23

Fig 3.21 Transitone Car Radio. 1937.

Fig 3.22 Philco advertisement, *Montreal Gazette*. 1937.

Fig 3.23 Miss Echo St. Pierre at the switchboard, Montreal, QC. 1945.

Fig 3.24

Fig 3.25

the conflict that solidified the crown corporation's place in Canadian homes.

Just as radio manufacturers had marketed their sets as providing access to political or cultural events in the past, new models were extolled for their ability to link listeners to events overseas. The RCA-Victor "Globe-Trotter" Model A-20, possibly designed by John Vassos, is a good example (Plate 3.14).[79] An advertisement simply presenting a photograph of the modern wooden receiver with angled, rectilinear dial, and two distinctive chrome elements bisecting the grill cloth over the speaker, reads "Get war news direct from Europe".[80] When paired with a map also produced by RCA Victor, indicating shortwave stations in Europe (Plate 3.15), the radio receiver provided access to the front. The sleek modern design of this model connotes efficiency and control, no doubt appealing to anxious Canadians desiring clear and accurate reports from Europe.

Many advertisements during the conflict highlighted manufacturers' contributions to the war effort and counselled restraint on the part of consumers. A prime example is a Rogers Majestic advertisement from 1942 (Fig. 3.24). "Seize the Radio Station," the copy reads below a depiction of saluting Nazi soldiers. Without question, Hitler used radio strategically before and during the war, including the staged attack on a station in Gleiwitz used as a justification to invade Poland.[81] Radio, normally cast as a benign invention offering freedom of travel, is here described as "a power for evil — a force to smash men's liberty." Like other manufacturers, Rogers Majestic announced its alignment with the Allied cause, especially since radio equipment was so important to modern warfare.

An advertisement for RCA Victor illustrated another way the seemingly invisible places of radio were essential to the war effort (Fig. 3.25). The main illustration represents a small, snow-covered structure linked to a radio antenna strung between two tall towers, while an aircraft darts across the northern sky. An inset shows a radio operator with headset and microphone broadcasting from this remote location. The copy not only underlines RCA Victor's vital contributions to the war effort, it also places radio in the vast Canadian territory:

> On Labrador's bleak coastline — in the frozen wastes of the regions rimming Hudson's Bay — along the rocky shores of British Columbia — Canada's radio outposts form a chain of listening posts — a ring of sentinels that maintain constant vigil to guard us against surprise attack. How great is the task of these outpost radio stations perform, can best be realized by visualizing 'the endless nowhere' of Canada's sparsely peopled Northern regions . . . visualizing the army of sentries we would need if each of the countless miles of Canada's coastlines had to be patrolled without the aid of radio's swift communication.

Radio stations are depicted as technologically advanced guard posts, maintaining and protecting the nation, just as Distant Early Warning radar installations would be in the Cold War.[82] The advertisement represents the vast Dominion of Canada as thoroughly modern, with the communications and transportation infrastructure running coast to coast (to coast) seen as essential to the nation state's composition as well as its security.

The CBC contributed to the erection of radio infrastructure during the war, most notably in the construction of its International Service station in Sackville, New Brunswick in 1944-45 (Fig. 3.26).[83] Like the other CBC stations, the corporation's architecture department designed the structure, although now in more "International Style" fashion rather than Art Deco, which was used in the earlier stations (including the station CBA opened on this site in 1939). The new station accommodated an existing 50kW standard broadcast transmitter on the lower level of the building, and added two new 50kW shortwave transmitters to the upper floor (Fig. 3.27). The choice of site on the Tantramar Marsh with its high mineral content meant that radio signals were even stronger. But most importantly, the design of antennas, hung like curtains across the

Fig 3.26

Fig 3.27

Fig 3.24 Rogers Majestic advertisement. *Montreal Gazette*. 1942.

Fig 3.25 RCA Victor advertisement. *Montreal Gazette*. 1942.

Fig 3.26 View of CBC International Service Transmission Station from tower in European array. Sackville, NB. 1945.

Fig 3.27 Main control room floor. CBC International Service Transmission Station. Sackville, NB. 1945.

towers and reaching heights of over 450 feet, allowed radio beams to be directed to almost anywhere in the world.[84]

Initially, CBC's International Service, broadcast from its Montreal studios, entertained Canadian troops and played a part in the propaganda effort, including broadcasting reports from German POWs. During the years following the conflict, the International Service continued its propagandist activities, celebrating the virtues of Canada and the Canadian way of life, in opposition to communism. During war time, radio installations were understood as a defence system and as part of a stealthy form of ideological warfare, which continued into the Cold War era.[85] The effective exercise of "soft power" through radio broadcasting required continual investment in infrastructure.

Building for Radio in Reconstruction Canada[86]

In the postwar era, the CBC resumed its prewar building campaign, opening, enlarging, or altering facilities across the country. By 1950, the Crown corporation had high-power transmitters in all of the western provinces (with two opening in Alberta and Manitoba), had expanded coverage in Ontario and Quebec, and had integrated the last provincial addition to Confederation, Newfoundland, into the national networks.[87] While studios had opened in the Hotel Vancouver in 1940 and plans were underway to provide new facilities in Winnipeg, the most exciting expansion of facilities for the CBC was the opening on May 18, 1951 of the Radio-Canada Building in the former Ford Hotel on Dorchester Street (now Boulevard René-Lévesque) in Montreal.[88]

This project came about after an explosion destroyed the CBC's main Montreal studios in January 1948. The corporation's engineering department surveyed existing buildings in the city and determined that, given its steel-frame structure and spacing of columns, the Ford Hotel (1930) was the only suitable location for the new facility.[89] Renamed the Radio-Canada Building, the remodelled 12-storey facility consolidated studios, technical and administrative services for both the domestic and international services, making it the central node in the national network. The opening, attended by dignitaries including the commissioners of the Royal Commission on National Development in the Arts, Letters, and Sciences, also showcased the CBC's new television studios.[90] While television would eventually alter the place of radio in Canadian homes, radio was still indispensable in the postwar welfare state.

Like the earlier CBC stations, the Radio-Canada Building provided a modern face for the corporation. And now, it was in Canada's largest city. The treatment of the ground floor lobby indicated an engagement with the latest trends in international modern architecture — full of light, open space, and reflective surfaces — as well as a

Fig 3.28

Fig 3.29

Fig 3.28 Main lobby. Radio-Canada Building. Montreal, QC. 1951.

Fig 3.29 Isometric Drawing. First Floor, Radio-Canada Building. Montreal, QC. 1951.

Fig 3.30

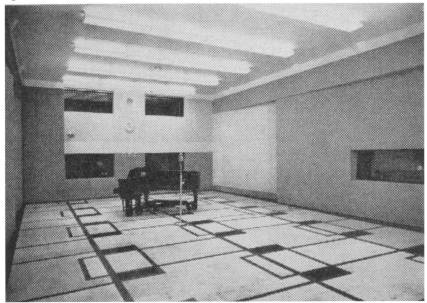

promenade architecturale leading visitors from the lobby to a mezzanine and upwards, past a window to a recording studio (Fig. 3.28). The first two floors contained the studios, while the upper floors held offices for newsrooms, record library, program staff, television department, administration personnel, the International Service, engineering, and, at the top, the board room for the Board of Governors and General Manager's office.[91] Like the earlier transmission stations, radio operations (in this case production) were staged for visitors, creating a physical, public space to complement the virtual space of radio experienced at home.

The media corporation unsurprisingly represented itself as modern and cutting edge, and indeed the interiors, with innovative acoustic treatment designed by McKinstry, would have appeared quite cosmopolitan in early 1950s Montreal. A press release from the CBC described some of the technical and aesthetic features, noting that "a different color scheme [was designed] in each of the 26 ultra-modern radio studios."[92] It goes on:

> Each studio was suspended on springs or rubber within the framework of the walls (as a box within a box). The inner studio walls, ceilings and floors are specially treated to preserve the required acoustics within the studios and the outer walls are similarly treated to isolate the studio from any outside noises. Within the studios themselves all acoustic tile has a different treatment on its reverse side. Thus, by

merely reversing a few tiles within the studio any one of a dozen different acoustical results can be achieved — making the Radio-Canada Building studios among the most versatile as well as most modern in North America.[93]

Fig 3.31

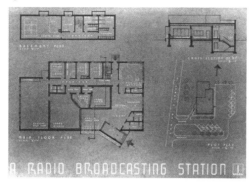

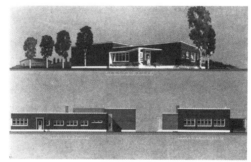

Visually, spatially, and acoustically, the Radio-Canada Building represented radio (and budding television) for visitors, and the careful design of the studios by the corporation's architects and engineers affected the experience of acoustic space for audiences (Fig. 3.30). For example, an orchestra playing in a studio could, through proper placement of microphones and musicians, be made to sound like a concert hall for radio listeners.[94]

The CBC was of course not the only producer of radio architecture in the country. The pages of the *The Royal Architectural Institute of Canada* contained references to radio facilities, including award-winning student projects of 1940.[95] These projects resonated quite closely with the design of the CBC's transmission stations, which had just opened. A few years later in 1947, the *JRAIC* published William G. Hames's student work on a radio building, indicating continued interest the form of radio architecture in the postwar era (Fig. 3.31). The splayed marquee with the "CBES" call sign letters address motorists on the highway. Again, transportation and communications infrastructure overlap, but now the highway has eclipsed the railway. According to the text, "[t]his station would serve as an advertising medium [for an association of Canadian Retail Merchants], as well as a cultural agent for the surrounding communities."[96]

This example not only highlights the intersection of modernism and radio architecture, but reminds us of the important place of private broadcasters in Canada.

Fig 3.30 Recording studio in Radio-Canada Building. Montreal, QC. 1951.

Fig 3.31 W.G. Hames' design for a radio station. *Journal, Royal Architectural Institute of Canada*. 1947.

Fig 3.32

Fig 3.34

Fig 3.32 Exterior view of CKWX Radio Station. Vancouver, BC. 1956.

Fig 3.33 CKWX Radio Station plan and section. Vancouver, BC. 1956.

Fig 3.34 CKWX Radio Station entry at night. Vancouver, BC. 1956.

During the Massey-Lévesque Commission consultations and afterwards, the Canadian Association of Broadcasters (representing private broadcasters) criticized the radio system in Canada for the fact that the CBC regulated all stations, yet, with its adoption of commercial material competed with private stations for advertising revenue.[97] It also contended that the private stations served communities better than the centralized, national networks. It is important to note that the CBC relied on private stations for the operation of its three networks. In 1947, for example, the CBC's Trans-Canada Network consisted of 24 basic and nine supplementary stations; and of the basic stations, only eight were owned by the CBC.[98] Its Dominion Network had only one station (CJBC Toronto) owned by the crown corporation. The additional 29 basic and 13 supplementary stations of this network, which ran lighter entertainment, were privately owned. And the French Network consisted of three basic stations owned by the CBC and ten supplementary ones.

Private stations were thus crucial to the built environment of radio and included some fine examples of modern architecture that embodied the social democratic ethos of the Reconstruction era.[99] An exemplar was Vancouver station CKWX designed by Ron Thom, who was working at Thompson, Berwick, Pratt, one of the leading architectural firms in the country (Fig. 3.32). While the origins of the CKWX date back to 1923, this was the first purpose-built structure for the powerful private station when it opened in 1956.[100] Located several blocks south of the newly opened 22-storey BC Electric Tower by the same architectural firm, the CKWX station is an excellent example of West Coast Modernism.[101]

Thom's design for the station responded well to its urban site near the Burrard Street Bridge (Fig. 3.33). A concrete core containing studios and control booths was surrounded by a ring of well-lit office space, a functional arrangement reminiscent of the design of the 1932 BBC Broadcasting House in London, UK.[102] Thom worked closely with CKWX engineer Charles Smith to master the acoustic requirements.[103] The corridor around the central studios was lit by a continuous skylight, brightening the space throughout the day. CKWX was ennobled with a mural by noted artist B.C. Binning. Binning's mural, described by architectural historian Rhodri Windsor Liscombe as "symbol[izing] the station's gathering and transmission functions in recurrent square and triangular patterns", enlivened the space and glistened at night under artificial lighting, providing a friendly gesture to the street (Fig. 3.34).[104] Binning had worked closely with Thom at University of British Columbia and on other architectural commissions, and developed a carefully crafted colour scheme in keeping with a West Coast palette.[105] Although on a much smaller scale than the CBC Radio-Canada Building in Montreal, the

Fig 3.33

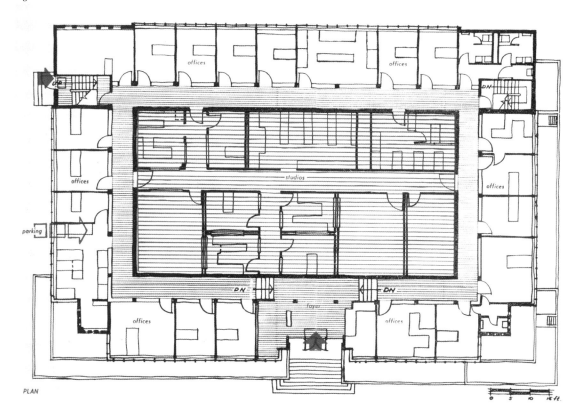

PLAN

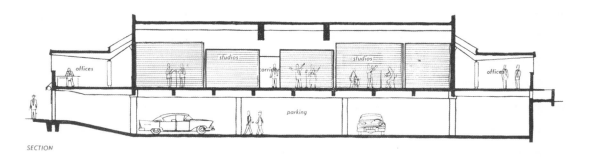

SECTION

design emphasized similar themes of transparency and openness, despite both buildings being located on busy streets and thus presenting challenges for recording/broadcasting.

The building was well received both critically and by the public. In 1958 the CKWX building won a Massey Silver Medal for architecture and in this same year the station was reaching an audience of 191,599.[106] F. H. Elphike, Vice President and General Manager of CKWX noted in a letter to B.C. Binning: "We are getting nothing but good comment on everything about the building, and this is after having some ten thousand people visit us."[107] The station was a leader in news coverage, had developed unique local programming, covered local sports, and thus played a role in the creation of local and regional radio publics.

In a discussion of the building in *Canadian Architect* in 1957, the author explained that the executives of CKWX "wanted something which would stand as a beacon of communication in the public eye (visually, as well as over the airwaves)."[108] Since private radio relies on sponsorship, its executives "were well aware that $250,000 spent on a fine building would be valuable advertising itself."[109] Just as William Hames predicted with his student project of ten years earlier, a well-designed radio station would serve as an advertisement for the station or corporation. The design of radio stations in the postwar era relied on the talent of architects to represent the medium of radio, the business of running it, and its promise for Canadian society.

Ubiquitous Radio

While radio ownership grew dramatically in the 1930s, its place in homes was cemented in the postwar period. In an article published in the CBC staff magazine *Radio* in 1949, H.F. Chevrier noted that there were 3,147,600 "radio homes" in Canada.[110] This meant that 94% of Canadian households had radios, resulting in an estimated 50% increase in listeners in the eight years since 1941. As a point of comparison, in 1949 only 50% of households had a telephone. Of course, these numbers were contingent on the erection of radio infrastructure, as discussed above, in addition to patterns of listening and the diminishing cost of radio receivers.[111] Joy Parr's research on postwar consumer culture in Canada reminds us that Canadians approached the market much more cautiously than did their southern neighbours (indeed the economic recovery was slower), and did not embark on the purchase of appliances as quickly.[112] She does not discuss the purchase of radios however, which were more affordable and seen perhaps, as more important in daily life than washing machines. For a generation that had known great uncertainty and instability, consumer purchases were tinged with moral values.[113] A radio allowed for broader participation in public culture. And while the television would eclipse the radio at the centre of the home,[114] the radio's portability allowed it to move into any room and almost anywhere outside. After being established as a lifeline to news events during the Second World War, in the post-war period access to radio continued to be an important part of Canadian life.

Radio provided a forum for nation-building, offering space for policy debates in the post-war era.[115] School broadcasts helped reinforce national values at a time of increased immigration, and new schools and libraries were built using modern design principles seen in some earlier radio architecture.[116] The suburban communities that grew up around these were predicated upon modern infrastructure such as electricity, telecommunications systems, and highways. A 1945 advertisement for Westinghouse, for example, boasted that the company was "selected by the designers of Thorncrest Village" in the Toronto-area suburb of Islington, for all household appliances including radio (Fig. 3.35). Still, even in the postwar era, a house full of large electric appliances remained a fantasy for many middle-class Canadians.[117]

Suburban but still connected, the home depicted here might remind of us of Marshall McLuhan's later work on auditory space. In 1961 he announced in *Canadian Architect* that, with the development and incorporation of electronic technology in Western society, "No place is a margin."[118] The instantaneousness of information from radios (and later televisions) meant that the suburbs were just as central to the experience of modernity as the metropolis. As radios migrated into automobiles and were carried as portables, the electronic medium can be seen as reinforcing the spatiality of the postwar suburb, that is, the omnipresent simultaneity of the suburban existence.

Northern Electric's popular "Baby Champ" models, launched in 1946, are a good example of the kind of radio proliferating in Canadian homes in the immediate postwar period (Plate 3.16). Designed during the war by engineers in Montreal, the streamlined receiver with a rainbow motif linking speaker to dial connoted mobility. It was lightweight and powerful, and came in six different colours (Plate 3.17). An example of postwar retooling, the "Baby Champ" captured the optimism of the period, as Canadians weary of years of economic hardship and rationing could finally re-enter the market. A General Electric brochure from 1944 stressed that company's focus was on satisfying consumer design preferences, exemplifying the promise of a postwar society in which individual needs and tastes would be satisfied by benevolent corporations (Plate 3.18).[119]

Radios like the "Baby Champ", or Addison's model A2A (Plate 3.19) were made of early plastics. With a wider range of colours available, affordable "miniatures" such as these allowed bedrooms, basements, and kitchens to become venues for listening for children, tinkering fathers, or busy homemakers.[120] The radio colours coordinated well

Fig 3.35

Fig 3.35 Westinghouse advertisement. *Canadian Homes and Gardens*. 1945.

Fig 3.36

Fig 3.37

with synthetic flooring materials such as "Marboleum". An advertisement for Armstrong's asphalt tiles, like the design of the mid-1940s plastic radios, also shows the continuity between the immediate pre and postwar design (Plate 3.20). An increasingly important feature of postwar houses designed for growing baby boom families, the basement became a key site of leisure and domestic activities. Positioned next to a lounge chair, the white plastic miniature model is imagined as the companion for mother while doing laundry, the father while at the workbench, or perhaps for someone looking for a quiet retreat.

As the postwar period unfolded, designs of mantel models evolved, influenced by other aspects of Canadian culture. The Crosley model E-15 of 1953 reflected the wider accessibility of automobiles in Canadian society (Plate 3.20). Perhaps connoting the freedom and mobility afforded by the car, this plastic radio looks like the front end of an automobile, with the speaker as a grill and symmetrically-placed dials as headlights. Of course, automobile listening also burgeoned in this period, eventually leading to subcultural listening practices of 1950s teenagers tuning into rock n' roll stations.[121]

An advertisement for Admiral radios emphasized "anti-inflationary prices" to allow for radios "in every room," as well as outdoors and "in your summer cottage" (Fig. 3.36). Not only does this ad underline the spatial flexibility of radio (that it *can be* in all of these spaces) but also the fact that it perhaps *should be* in every room, as part of a modern lifestyle. Cottage lifestyle developed in the postwar period as roads improved and the Canadian middle class became more affluent. For example, an article in *Canadian Homes and Gardens* from July 1949 offered "10 ways to pep up your cottage".[122] One of the ways listed was for do-it-yourselfers to integrate earlier radios into modern, multifunctional furniture. The pre-war highboy models were definitely out of fashion, even at the cottage.

While portable radios date back to the beginning of the technology, they were rather cumbersome until the late 1930s.[123] The RCA Victor P31 from the early part of that decade, for instance, was basically a suitcase containing a radio receiver (Plate 3.21). According to historian Michael Brian Schiffer, the turning point in the history of portable radios in the United States occurred in 1938, with the development of smaller "low-drain" battery tubes. An example from this moment is the Marconi portable released just before the war (Fig. 3.37). Enclosed in a waterproof case, it was about the size of a lunch box, and indicated one way that radio would develop after the war.

Indeed the market for portable radios really broadened after the global conflict.[124] Smaller vacuum tubes and eventually transistors in the 1950s, combined with an appetite for mobile listening, helped spark the growth of portable radio in the postwar period. The Sparton "Hiker" of 1948 provides an example of the new portable tube radios (Plate 3.22). Designed with faux alligator skin case, the radio is depicted in a contemporary advertisement worn by a woman like a purse or perhaps camera bag. (Fig. 3.38). The advertisement expresses the urge to take radio out into the world, including while camping. The Hallicrafters 1953 model "Continental 5R40" is another example of a postwar portable radio (Plate 3.23). The artefact held by the Canada Science and Technology Museum collection has been well used, and is marked with an Ottawa Roughriders sticker (on its top and thus not visible in the image), suggesting it may have been brought to football games and not just used at home. The "lunch box" style of portable tube radios continued through the mid-1950s, until gradually succeeded by transistor models after that.[125] The RCA Victor P338, from 1959, exemplifies the new, lightweight, plastic transistor models (Plate 3.24). Portability breathed new life into radio, as television supplanted its role as electronic hearth at home.

Radio in the postwar period became increasingly part of an enlarged entertainment area. The idea of a "musicorner"— an integrated space for phonograph and radio, and later television — was developed by John Vassos at RCA and put on display in the 1939-40 New York World's Fair.[126] RCA continued to sell the idea of the "musicorner" in the postwar era, advising readers of *Canadian Homes and Gardens* in 1949 to "Save One Corner for Music."[127] Owners especially of small homes could build in storage for records and equipment to maximize their living room's footprint. A year later, we see an entertainment corner built into a Thompson, Berwick, and Pratt house for Bob and Helen Berwick (Fig. 3.41). For a custom house like this, radio could be considered standard equipment. However, by the mid-1950s, radio was less of a focal point and more likely to be part of a modern entertainment centre. The cover of *Canadian Homes and Gardens* from 1955 shows how well radio fit into modern furniture in the postwar home (Plate 3.25).

In fact in this era, radio could fit into just about any piece of furniture with relative ease. An Electrohome pamphlet from 1947 features "Rhapsody" furniture to house television, radio, or phonograph, providing a good example of how new media technologies could be accommodated into the style of any home (Plate 3.26). Occasional tables made by Deilcraft underlined the contemporary, more informal use of the living room that included "coffee tables" as well as state-of-the art entertainment. Design historian Judy Attfield reminds us how significant certain furniture items, such as the coffee table, were to social practice and domestic interiors at mid-century.[128] The coffee table, she argues, signalled a movement toward less formal (and more modern)

Fig 3.38

Fig 3.36 Admiral's advertisement. *Montreal Gazette*. 1948.

Fig 3.37 Marconi advertisement. *Montreal Gazette*. 1939.

Fig 3.38 Sparton advertisement. *Montreal Gazette*. 1948.

Fig 3.39

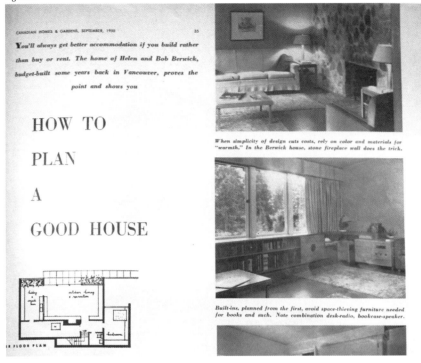

social interaction in the U.K. In this book, we have made a similar argument for the influence of radio – as medium and object – on domestic spaces and social practices.

A more extreme version of the open, informal, and casual lifestyle at mid-century in Canada was featured in *Canadian Homes and Gardens* in 1956 (Fig. 3.39). Designed by Ottawa-based architect Jim Strutt, the plan of the Dawson House in Senneville, Quebec is organized around two hexagonal shapes in an open-concept plan. At the intersection of the children's "wing" to the right and the "adult" areas to the left is a radio, a record player, and a television set. The built-in radio sits near the fireplace, as it had in the ideal floor plans of the early 1930s, but it is now also at the centre of the home. Of course, in a such a uniquely designed home, placing the radio in the centre of the plan makes sense acoustically, but in some ways the plan also underlines the centrality of radio in this period. The form of the space was, at least in part, designed around modern, electro-acoustic soundscapes and lifestyles that included deeply entrenched patterns of radio listening.

Fig 3.40

Despite the rise of television, radio was a key player in postwar Canadian social space. From its place in each home, it continued to offer a conduit to a broader public discourse. Aligned with social democratic ideals of universal education and access to information, radio became constitutive of modern Canadian citizenship. It was centrally located in many houses and more ubiquitous than ever, providing companionship, news, and entertainment for Canadians wherever they were.

By the Cold War, access to radio was a given. While the CBC's International Service created a representational space of Canada abroad, domestically both public and private stations contributed to the shape of Canadian social life. Playing in the background of everyday life, radio wrought significant changes in Canadian homes, domestic life, and the public arena, as it shifted perceptions of space and society, and perhaps even what it meant to be a modern Canadian citizen.

Fig 3.39 Article in *Canadian Homes and Gardens*. 1950.

Fig 3.40 Article in *Canadian Homes and Gardens*. 1956.

Plate 3.1 Westinghouse Radiola III, 1924.

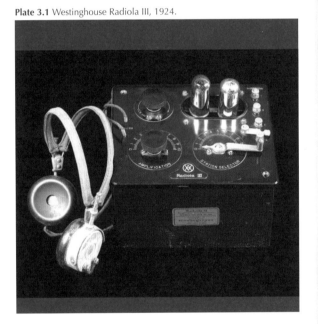

Plate 3.2 Cover of Eaton's *Radio Catalogue*, 1924-25.

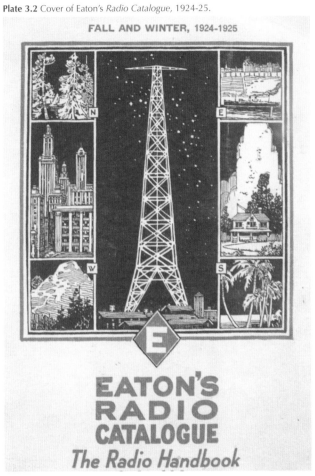

Plate 3.3 Cover of Eaton's *Radio Catalogue*, 1925.

Plate 3.4 Back cover of Eaton's *Radio Catalogue*, 1925.

Plate 3.5 Rogers Batteryless Model 110, 1925-1926.

Plate 3.6 Stromberg-Carlson Model 25-A. 1932.

Plate 3.7 Philco 20. 1930.

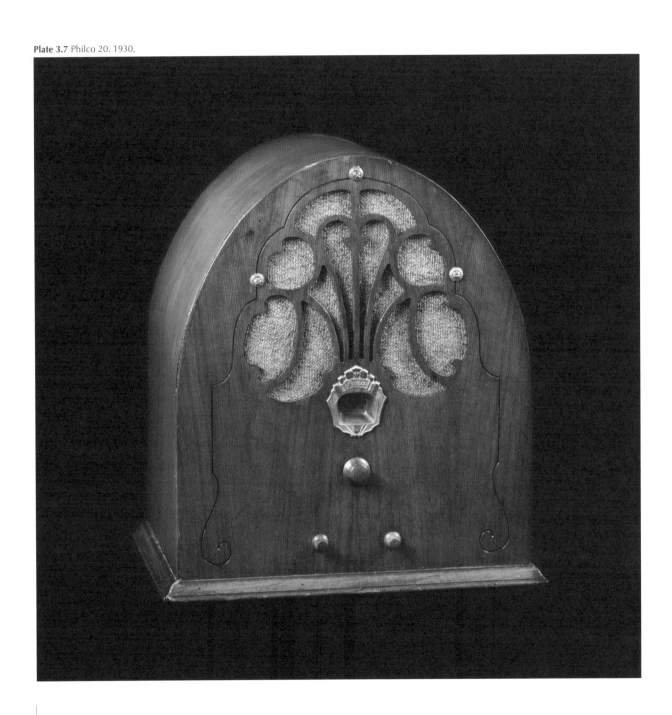

Plate 3.8 Westinghouse "Columaire" (Model 801). 1931.

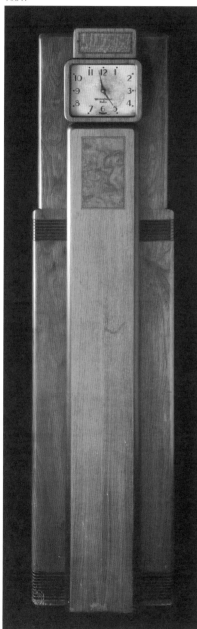

Plate 3.9 Dominion Inlaid Linoleum advertisement. *Maclean's Magazine*. 1936.

Plate 3.10 RCA Globetrotter C-6-1. 1935.

Plate 3.11 Front cover of Eaton's *Radio Catalogue*. 1935.

Plate 3.12 General Electric Radiotrons Map. 1935.

Plate 3.13 RCA Victor (Model A-1). 1939.

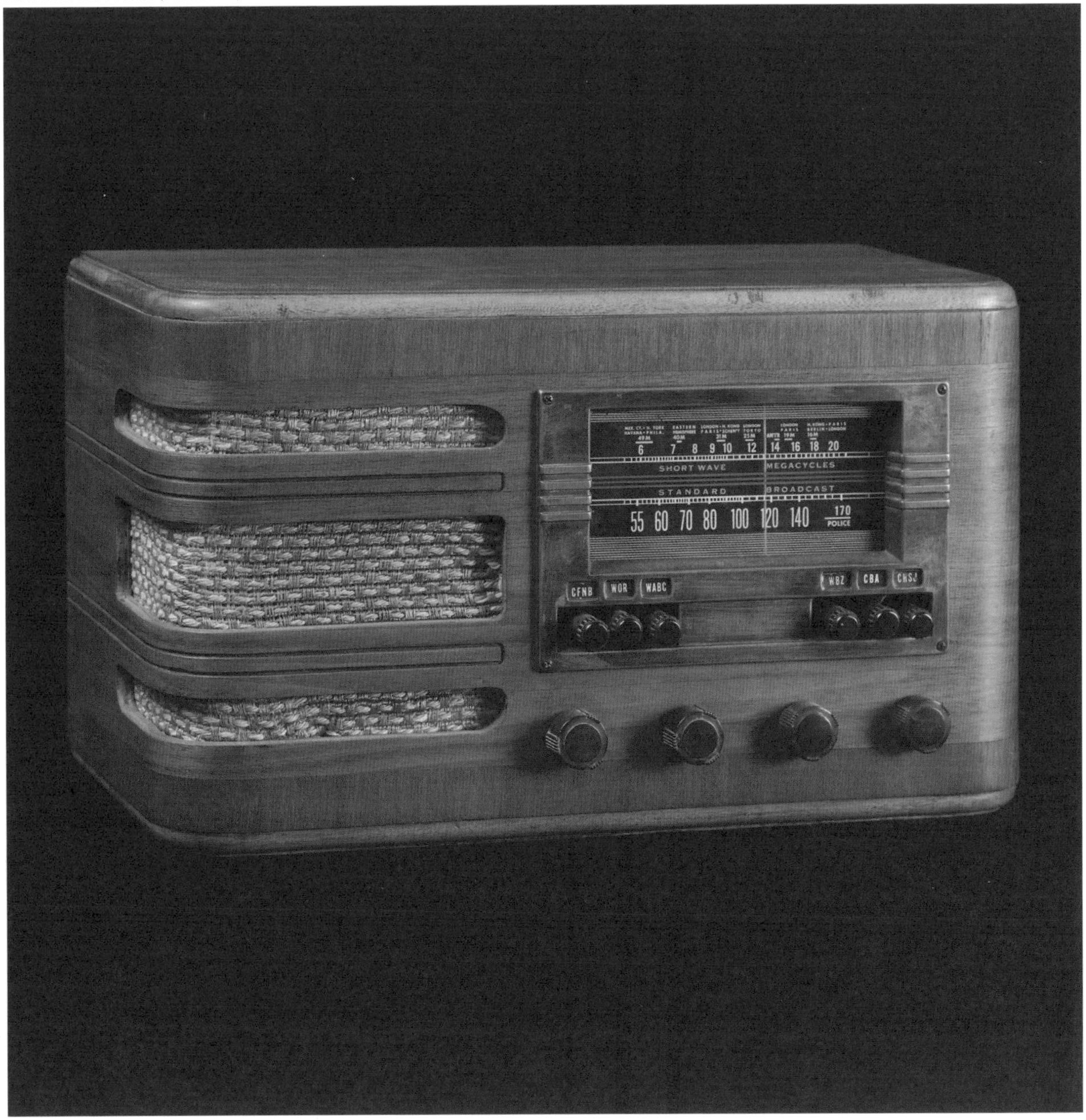

Plate 3.14 RCA Victor A-20. 1940.

Plate 3.15 RCA Victor Radio Map. 1940.

Plate 3.16 Northern Electric "Baby Champ" (Model 5110). 1946.

Plate 3.17 Northern Electric advertisement. *Chatelaine*. 1947.

Plate 3.18 General Electric pamphlet. 1944.

Plate 3.19 Addison A2A. 1945.

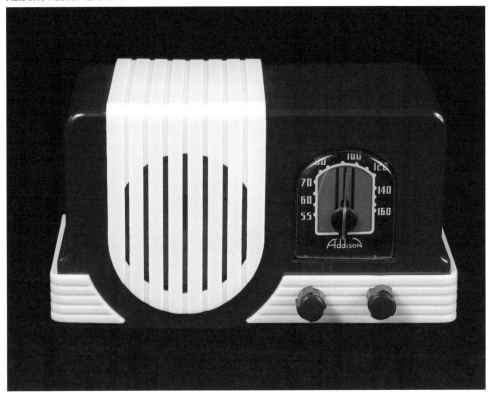

Plate 3.20 Armstrong advertisement. *Canadian Homes and Gardens*. 1949.

Plate 3.21 Crosley E-15. 1953.

Plate 3.22 RCA Victor (Model P31). 1931-1932.

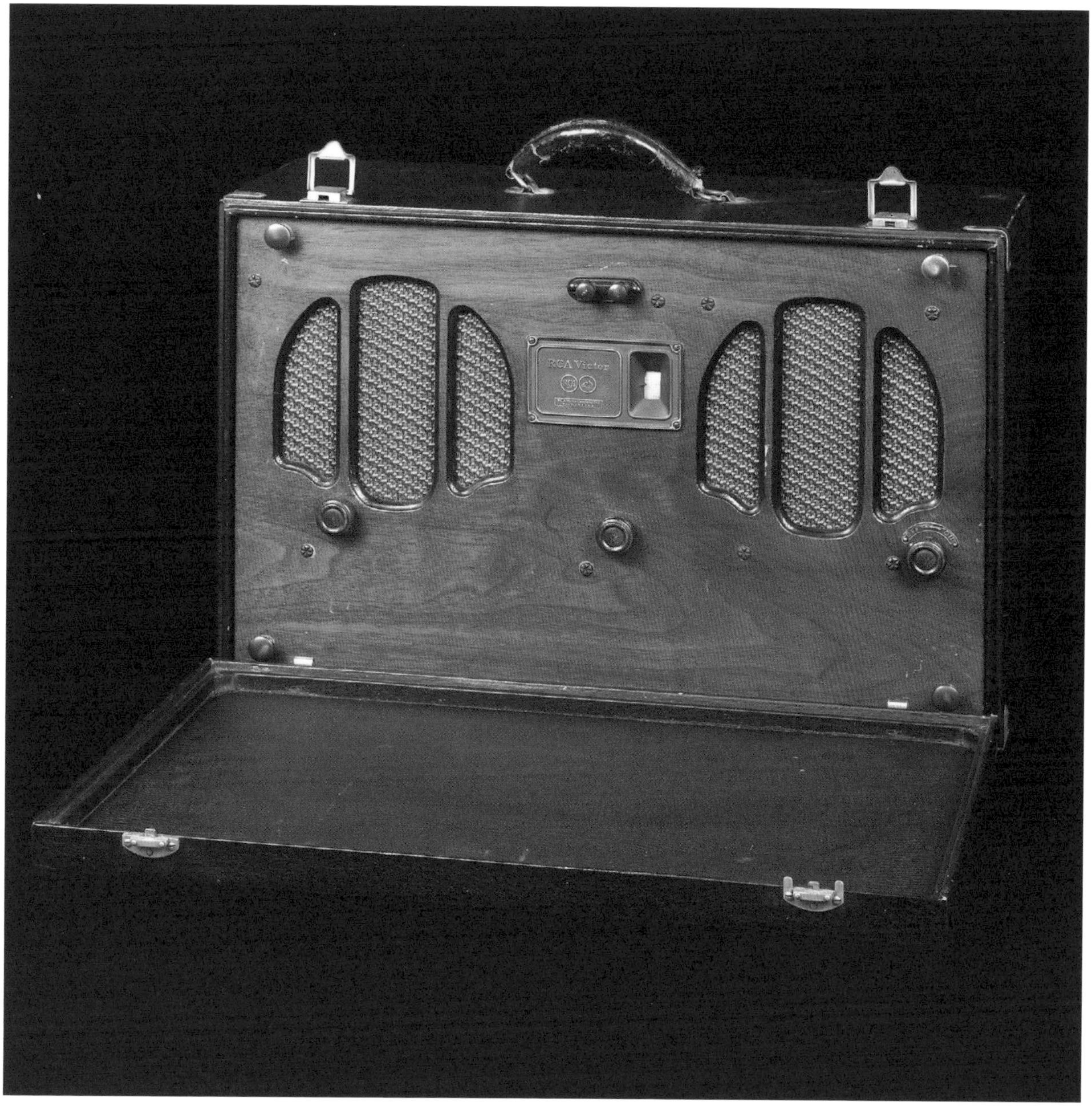

Plate 3.23 Sparton Hiker. 1948.

Plate 3.24 Hallicrafters Canada, Continental 5R40. 1953.

Plate 3.25 RCA Victor (Model P-338). 1959.

Plate 3.26 Cover of *Canadian Homes and Gardens*. 1955.

Plate 3.27 Electrohome pamphlet. 1947.

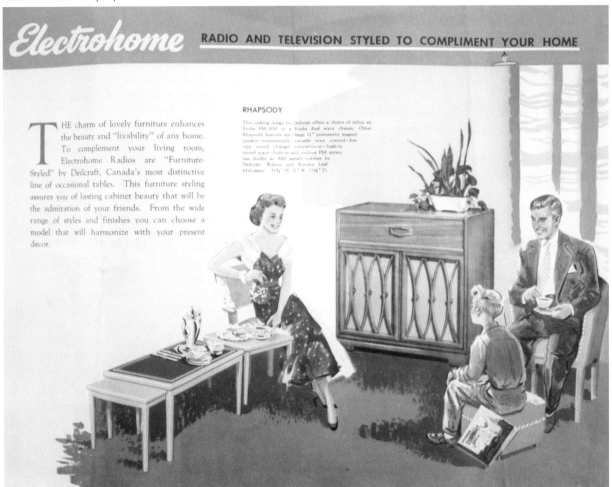

Notes

[1] Space is considered in dynamic terms in this chapter, as incorporating perceived, conceived, and lived spaces, and is thus influenced by Henri Lefebvre, *The Production of Space*, trans. Donald Nicholson-Smith (Oxford and Cambridge: Blackwell, 1991).

[2] This is discussed in "Listening to Deco: Sound Design in Canada," in Michael Windover, *Art Deco: A Mode of Mobility* (Québec: Presses de l'Université du Québec, 2012), 246-247. For more on theorizations of media space, see Jesper Falkheimer and André Jansson, eds, *Geographies of Communication: The Spatial Turn in Media Studies* (Göteborg: Nordicom, 2006).

[3] See Jason Loviglio, *Radio's Intimate Public: Network Broadcasting and Mass-Mediated Democracy* (Minneapolis: Minnesota University Press, 2005) for an in-depth exploration of this idea in the American context and Len Kuffert, "'What Do You Expect of This Friend?': Canadian Radio and the Intimacy of Broadcasting," *Media History 15*, no. 3 (2009): 303-317. Kuffert expanded on this in his recent book, *Canada Before Television: Radio, Taste, and the Struggle for Cultural Democracy* (Montreal & Kingston: McGill-Queen's University Press, 2016), especially chapter one.

[4] See Michael Windover, "Designing Public Radio in Canada," *RACAR* 40, no. 2 (2015): 42-56.

[5] Kate Lacey, *Listening Publics: The Politics and Experience of Listening in the Media Age* (Cambridge: Polity Press, 2013). On the discursive formation of publics, see Michael Warner, "Publics and Counterpublics (abbreviated version)," *Quarterly Journal of Speech 88*, no. 4 (November 2002): 413- 425.

[6] The American chain, Hotels Statler, announced the incorporation of "Radio in Every Room" of their six hotels (in Boston, Buffalo, Cleveland, Detroit, St. Louis, New York) in an advertising campaign printed in *MacLean's Magazine* in 1928. A sign of radio's domesticity is highlighted in the advertisement printed in the March 15 issue which reads "Statler makes hotels more homelike…with…radio in every room!" p. 57.

[7] George H. Buck, "The First Wave: The Beginnings of Radio in Canadian Distance Education," *Journal of Distance Education* 21, no. 1 (Spring 2006): 75-88. For more on CNR's radio service, see E. Austin Weir, *The Struggle for National Broadcasting in Canada* (Toronto: McClelland and Stewart, 1969), Frank W. Peers, *The Politics of Canadian Broadcasting*, 1920-51 (Toronto: University of Toronto Press), 22-27, and Mary Vipond, *Listening In: The First Decade of Canadian Broadcasting*, 1922-1932 (Montreal and Kingston: McGill-Queen's University Press, 1992), 50-51, and Michael Windover, "Transmitting Nation: 'Bordering' and the Architecture of the CBC in the 1930s," *Journal of the Society for the Study of Architecture in Canada* 36, no. 2 (2011): 5-12.

[8] Buck, 79.

[9] The Canadian Radio Broadcasting Commission acquired stations CNRO (Ottawa), CNRA (Moncton), and CNRV (Vancouver), as well as studios in Montreal and Halifax from the CNR for $50,000 on Mar. 1, 1933. See Weir, 139-141.

[10] See Harold A. Innis, *A History of the Canadian Pacific Railway* (Toronto: McClelland and Stewart, 1923). For more on Innis' late work and its impact on historiography, see Graeme Patterson, "Harold Innis and the Interpretation of History," in *History and Communications: Harold Innis, Marshall McLuhan, the Interpretation of History* (Toronto: University of Toronto Press, 1990), 1-22. Innis' important late works include *The Bias of Communication*, foreword by Marshall McLuhan (Toronto: University of Toronto Press, 1951) and *Empire and Communications* (Oxford: Clarendon Press, 1950).

[11] See Marshall McLuhan, *The Gutenberg Galaxy: The Making of Typographic Man* (Toronto: University of Toronto Press, 1962). Benedict Anderson argues a similar point about the role media (especially print) play in the creation of nation states in *Imagined Communities: Reflections on the Origin and Spread of Nationalism* (London: Verso, 1983). For one application of Anderson's ideas to radio, see Michele Hilmes, *Radio Voices: American Broadcasting*, 1922-1952 (Minneapolis: University of Minnsata Press, 1997).

[12] Richard Cavell argues that space was a crucial focus in McLuhan's work. See Richard Cavell, *McLuhan in Space: A Cultural Geography* (Toronto: University of Toronto Press, 2002). For more on Innis's critique of cultural imperialism, especially in regards to the role of the press, see his *Changing Concepts of Time* (Toronto: University of Toronto Press, 1952).

[13] For more on broadcasting content on the built environment within the British context, see Shundana Yusaf, Broadcasting Buildings – Architecture on the Wireless, 1927-1945 (Cambridge and London: MIT Press, 2014).

[14] Eric Wenaas argues that the Radiola III was a joint venture between Westinghouse and RCA in *Radiola: The Golden Age of RCA, 1919-1929* (Chandler, AZ: Sonoran, 2007), 441. A nearly identical model (although with different interior components) was produced by Canadian General Electric also in 1924.

[15] The Westinghouse Radiola III advertisement from the *Winnipeg Free Press* of December 13, 1924, Vol. 50, No. 138, p. 28 lists the price at $45 compared to the larger, 4-tube Radiola III-A at $80, or the $325 Radiola X, which included a built-in loudspeaker. None of these prices included the cost of batteries or antennas.

[16] Canadian Westinghouse Company, Limited. September 5, 1924. "The New Radiola — What everyone has been waiting for" advertisement. *Manitoba Free Press*, 9; Canadian Westinghouse Company, Limited. September 15, 1924. "Radiola Sets Now Ready" advertisement. *MacLean's Magazine*, 41.

[17] For more on the technological developments in radio reception in Canada, see Sharon A. Babaian, *Radio Communication in Canada: A Historical and Technological Survey* (Ottawa: National Museum of Science and Technology, 1992), especially chapters 8 and 9.

[18] For examples of this type of alternate listening practice, see Shaun Moores, "'The Box on the Dresser': Memories of Early Radio and Everyday Life," *Media, Culture and Society* 10 (1998): 30. See also Thomas Michael Everrett, "Ears Wide Shut: Headphones and Moral Design" (PhD diss., Carleton University, 2014), 58.

[19] For more on DXing in the US context, see Susan J. Douglas, *Listening In: Radio and the American Imagination* (Minneapolis and London: University of Minnesota Press, 2004), 57-58; Richard Butsch, *The Making of American Audiences: From Stage to Television, 1750-1990* (Cambridge: Cambridge University Press, 2000), 177-178.

[20] Everrett, 53-54.

[21] Everrett, 54 and images on 57.

[22] See Lynn Spigel, *Make Room for TV: Television and the Family Ideal in Postwar America* (Chicago and London: University of Chicago Press, 1992), 28. See also Everrett, 61-64.

[23] For more examples, see Windover, "Designing Public Radio in Canada," 48, which looks at the covers from 1936-37, and Michael Windover, "Placing Radio in Sackville, New Brunswick," *Buildings & Landscapes* 24, no. 1 (Spring 2017) (forthcoming), for a discussion of the 1926-27 covers.

[24] Everrett, 60.

[25] Everrett explains how the cone loudspeaker works: "In this new design, the metal diaphragm of the telephone receiver was replaced with a metal coil which, instead of moving in and out (toward the magnets), moved up and down (between the magnets). As the coil moved up and down, it caused a thin and wide paper diaphragm—attached to one end of the coil and outside of the magnetic field—to move in tandem. Because there was no longer a concern with touching the magnets ... this meant the coil could move in and out with greater force and without ever distorting the sound. And because the paper diaphragm was far larger than a telephone diaphragm, it could also move a lot more air, meaning that a horn was no longer required to amplify the sound" (66).

[26] See Windover, *Art Deco*, 208-211. For more on the invention of the A/C tube and the development of Rogers as a radio company, see Ian Anthony, *Radio Wizard: Edward Samuel Rogers and the Revolution of Communications* (Toronto: Gage Publishing for Rogers, 2000).

[27] See Shaun Moores, "'The Box on the Dresser', 23-40; see also Anne F. MacLennan, "Women, Radio Broadcasting and the Depression: A 'Captive' Audience from Household Hints to Story Time and Serials," *Women's Studies* 37, no. 6 (2008): 616-633, and her "Learning to Listen: Developing the Canadian Radio Audience in the 1930s," *Journal of Radio & Audio Media* 20, no. 2 (2013): 311-326. This is similar to the experience in the United States. See Susan J. Douglas, *Listening In*, Richard Butsch, "Crystal Sets and Scarf-Pin Radios: Gender, Technology and the Construction of American Radio Listening in the 1920s," *Media, Culture & Society* 20, no. 4 (1998): 557-572, and William Boddy, "Archaeologies of Electronic Vision and the Gendered Spectator," *Screen* 35, no. 2 (1994):105-122.

[28] Superheterodyne models did not require an external antenna.

[29] It should be remembered that player pianos and later gramophones and phonographs were also pieces of musical furniture. See Kyle S. Barnett, "Furniture Music: The Phonograph as Furniture, 1900-1930," *Journal of Popular Music Studies* 18, no. 3 (2006): 301-324.

[30] It was also expensive at $239 in the depths of the Great Depression. See Stromberg-Carlson. February 16, 1932. "A Name Synonymous with radio quality" advertisement. *The Montreal Gazette*, 2.

[31] Many thanks to Prof. Jill Carrick who suggested this iconographical reference.

[32] See Philco Products Ltd. July 19, 1932. "Attend the Opening of the Imperial Economic Conference with a 1933 Philco," advertisement. *The Montreal Gazette*, 4.

[33] See Adrian Forty's discussion of radio design in *Objects of Desire: Design and Society since 1750* (London: Thames and Hudson, 1986), 11-12.

[34] "The New Marconi is a decorative asset to any room," *Toronto Globe and Mail*, 31 October 1930, 5. On the early part of her career, see Gertrude Pringle, "Making a Business of Good Taste," *MacLean's Magazine*, 1 October 1926, 71-72. For an example of her prose, see Minerva Elliot, "New Fads Versus Old Traditions," *Canadian Homes and Gardens*, February, 1928, 32, 72. Her interior decorating of the home of Douglas Hallam (Toronto) was featured in *Canadian Homes and Gardens*, February 1930, 30-31.

[35] Whitney Dill, "We Furnish a Four-Roomed Apartment," *Canadian Homes and Gardens*, October- November 1934, 58.

[36] The Fleming-Bell Manufacturing Company went so far as to incorporate a radio into a cabinet in the shape of a baby grand piano. See Model B.C. in Lloyd Swackhammer, *Radios of Canada* (Alma, ON: Cober Printing Ltd, 2002). We should also keep in mind the place of the early electric intrusion into the parlor: the player piano.

[37] Philco Products Ltd. October 3, 1930. "Music to Match your mood! 4 Shades of tone, Brilliant-Bright- Mellow-Deep. With the new Philco," advertisement. *Vancouver Sun*, 3.

[38] See Windover, *Art Deco*, 218.

[39] See MacLennan, "Learning to Listen," and MacLennan, "Women, Radio Broadcasting and the Depression".

[40] Philco Products Ltd. October 11, 1930. "The Sensation of the Industry: Philco Baby Grand Radio," advertisement. *Manitoba Free Press*; 6; also, November 8, 1930. "If you already own a radio here's your chance to own a second set, one upstairs, one down or as an additional set for guest room or 'den. PHILCO BABY GRAND RADIO: The Wonder Set," advertisement. *Manitoba Free Press*; 20.

[41] See Windover, *Art Deco*.

[42] See Windover, *Art Deco*, 211-212.

[43] Canadian Marconi Company. September 8, 1930. "In this Modern New Plant Marconi has perfected Radio 3* Different Ways," advertisement. *The Montreal Gazette*, 11.

[44] See Windover, *Art Deco*, 221-237.

[45] See Adrian Forty, "Wireless Style: Symbolic Design and the English Radio Cabinet, 1929-1933," *Architectural Association Quarterly* 4 (Spring 1974): 23-31.

46 See Windover, *Art Deco*, 250.

47 Canadian Westinghouse Company, Limited. August 26, 1931. "See and Hear Columaire '8' Today," advertisement. *Globe and Mail*, 14.

48 See Windover, "Listening to Deco" in *Art Deco*. See also Christopher Long, *Paul T. Frankl and Modern American Design* (New Haven and London: Yale University Press, 2007).

49 See MacLennan, "Women, Radio Broadcasting and the Depression."

50 In *The Production of Space*, Lefebvre employs a dialectical triad of spatial practice (e.g., daily life activities), representations of space (e.g., the operational codes of space), and representational spaces (e.g., space as simply described or represented in art that challenges spatial conventions). See Lefebvre, 33 and 38-39.

51 And we should keep in mind that contemporary architects, planners, and designers were fascinated with time-space relations in the interwar period. One of the most influential considerations of this quality in modern architecture is Sigfried Giedion, *Time, Space, Architecture: The Growth of a New Tradition* (Cambridge: Harvard University Press, 1941).

52 See Patricia Cogdell, *Eugenic Design: Streamlining America in the 1930s* (Philadelphia, University of Pennsylvania Press, 2004).

53 Joy Parr's research shows that key appliances, such as washing machines and even stoves, were later additions to Canadian homes compared to the United States. See Joy Parr, *Domestic Goods: The Material, the Moral, and the Economic in the Postwar Years* (Toronto: University of Toronto Press, 1999).

54 See Windover, "Transmitting Nation," "Designing Public Radio in Canada," and "Placing Radio in Sackville, New Brunswick."

55 See Windover, "Transmitting Nation."

56 Mackenzie Waters, "Broadcasting Stations," *Journal of the Royal Architectural Institute of Canada* 15, no. 10 (1938): 215-218; Windover, "Transmitting Nation," 8.

57 See Gabrielle Esperdy, *Modernizing Main Street: Architecture and Consumer Culture in the New Deal* (Chicago and London: University of Chicago Press, 2008).

58 See Windover, "Transmitting Nation," 10-11.

59 The floor at CBK Watrous also contained a map. See Windover, "Designing Public Radio in Canada," 52-55.

60 Windover, "Transmitting Nation," 11.

61 See Windover, *Art Deco*, 243-245.

62 Foucault initially discusses heterotopias in discursive terms in the preface to *The Order of Things: An Archaeology of the Human Sciences*, Routledge Classics (London and New York: Routledge, 2002), xvi-xxvi. He delivered a radio address on "Utopia and Literature" on December 7, 1966, in which he described heterotopias as real places. He would later develop this concept in a lecture, "Of Other Spaces," delivered to the Circle of Architectural Studies in Paris on March 14, 1967. For more on this, see Daniel Defert, "Foucault, Space, and the Architects," in *Politics/Poetics: Documenta X—The Book* (Ostfildern-Ruit: Cantz Verlag, 1997): 274-283.

63 Michel Foucault, "Of Other Spaces," trans. Jay Miskowiec. *Diacritics* 16, no.1 (Spring 1986): 24. He then goes on to outline six principles of heterotopias complete with vastly different but concrete examples. Briefly, he notes that: 1) probably every culture has heterotopias to manage crisis or deviation, but they exist in different forms; 2) heterotopias can function differently over time to meet societal needs; 3) that a heterotopia can juxtapose several, sometimes incompatible spaces in a single place; 4) heterotopias are often linked to slices of time (or, what he calls, heterochronies); 5) that heterotopias "always presuppose a system of opening and closing that both isolates them and makes them penetrable"; 6) and finally, that they have a function in relation to all other space, either as an illusion that exposes all other spaces as being as illusory, or as an ideal space that contrasts with

the messiness of society.

[64] On the value of heterotopias for architecture, see Henry Urbach, "Writing Architectural Heterotopia," *The Journal of Architecture* 3 (winter 1998): 347-254.

[65] See Windover, *Art Deco*, chapter 4.

[66] Foucault, "Of Other Spaces," 27.

[67] For more on the role of imagination in global modernity, see Arjun Appadurai, *Modernity at Large: Cultural Dimensions of Globalization* (Minneapolis: University of Minnesota Press, 1996).

[68] Marshall McLuhan, *Understanding Media: The Extensions of Man* (New York: McGraw-Hill, 1964).

[69] See Windover, "Placing Radio in Sackville, New Brunswick."

[70] For more on the unveiling of television in 1939-40 and the design of RCA's TRK-12, see Danielle Shapiro, *John Vassos: Industrial Design for Modern Life* (Minneapolis and London: University of Minnesota Press, 2016), 165-182.

[71] Commenting on the American context, Butsch notes that as early as 1933 "the *New York Times* identified a 'craze' for car radios and attributed it to young people" (*The Making of American Audiences*, 205).

[72] For example, see J.J.H. McLean & Co. Ltd. April 14, 1934. "The 'Portette' Victor's 2 in 1 Auto Radio," advertisement. *Winnipeg Free Press*, 16.

[73] Butsch, *The Making of American Audiences*, 206. See also the chart on 176. More research needs to be done on the Canadian situation.

[74] Karin Bijsterveld, "Acoustic Cocooning: How the Car Became a Place to Unwind," *Senses & Society* 5, no. 2 (2010):189-211, especially 194-195.

[75] This phrase is used in Karin Bijsterveld, Eefje Cleophas, Stefan Krebs, and Gijs Mom, *Sound and Safe: A History of Listening Behind the Wheel* (New York: Oxford University Press, 2014). Their fascinating study explores the sonic design and auditory culture of automobiles, with a focus on Europe and the US.

[76] Windover, "Designing Public Radio in Canada," 49.

[77] For more on the architecture of the CBC, see Windover, "Transmitting Nation," "Designing Public Radio in Canada," and "Placing Radio in Sackville, New Brunswick."

[78] The building containing this interior is not identified in the issue of *Radio: CBC Staff Magazine*, although it is likely the King's Hall building in Montreal.

[79] For more on Vassos and his design career at RCA, see Shapiro, *John Vassos*, 81-108.

[80] Willis & Company Limited. April 16, 1940. "At Willis This New RCA-Victor Radio Globe-Trotter," advertisement. *The Montreal Gazette*, 2.

[81] Windover, "Transmitting Nation," 11 and "Placing Radio in Sackville, New Brunswick."

[82] For a discussion of the heritage of the Cold War northern radar installations, see Andrew Waldron, "Canada's Distant Early Warning Line," *Docomomo Journal* 38 (Mar. 2008): 56-60.

[83] For more, see Windover, "Placing Radio in Sackville, New Brunswick."

[84] Stuart Griffiths, "International Shortwave Broadcasting in Canada," *Canadian Geographical Journal* 33, no. 5 (November 1946): 218-35

[85] For more on the CBC's International Service, see James L. Hall, *Radio Canada International: Voice of a Middle Power* (East Lansing: Michigan State University Press, 1997) and Arthur Siegel, *Radio Canada International: History and Development* (Oakville, Ontario: Mosaic Press, 1996).

[86] For more on the culture of Reconstruction generally, see Nancy Christie and Michael Gauvreau, eds., *Cultures of Citizenship in Post-war Canada, 1940-1955* (Montreal and Kingston: McGill-Queen's University Press, 2003) and Leonard B. Kuffert, *A Great Duty: Canadian Responses to Modern Life and Mass Culture, 1939-1967* (Montreal and Kingston: McGill-Queen's University Press, 2003). For more on Canadian architecture in the Reconstruction era, see Harold Kalman, *A Concise History of Canadian Architecture* (Oxford: Oxford University Press, 2000), 535-568, and Rhodri Windsor Liscombe, *Architecture and the Canadian Fabric* (Vancouver: UBC Press, 2011), especially part 4 "Reconstructing Canada."

[87] The building projects are noted in the "development" or "engineering division" portion of the annual reports of the CBC. See the reports from 1946-51.

[88] See CBC annual reports 1950-51 and 1951-52.

[89] Canadian Broadcasting Corporation Annual Report, 1948-49 (Ottawa: Canadian Broadcasting Corporation, 1949), 43.

[90] See "Invitation List for the Official Opening of the Radio Canada Building, May 18 th , 1951," Library and Archives Canada, Canadian Broadcasting Corporation Fonds (hereafter LAC, CBC Fonds), RG41 Vol. 94, file 3-16- 14. See also the published booklet describing the features of the building in the same file.

[91] A document in LAC, CBC Fonds, RG41 Vol. 94, file 3-16- 14, describes the interior spaces of the CBC Building in the tone of a tour guide.

[92] "Radio-Canada Building Opens in Montreal—Latest Link in CBC's World-Wide Radio System," 4, LAC, CBC Fonds, RG41 Vol. 94, file 3-16- 14.

[93] "Radio-Canada Building Opens in Montreal," 4.

[94] See the discussion under "Studio 11 or 12" in what appears to be a tour script, LAC, CBC Fonds, RG41 Vol. 94, file 3-16- 14.

[95] "Report of the Jury of Awards, R.A.I.C. Competition, 1940," *Journal, the Royal Architectural Institute of Canada* 17, no. 3 (March 1940): 41-44, especially 42-43.

[96] W.G. Hames, "A Radio Broadcasting Station," *Journal, the Royal Architectural Institute of Canada* 24, no. 5 (May 1947): 146. Another evocative design from the University of Manitoba a few years later was also published in the *JRAIC*. See Ismay W. Haines, "A Prairie Regional Broadcasting Centre," *Journal, the Royal Architectural Institute of Canada* 27, no. 4 (April 1950): 142.

[97] For more on the Massey Commission, see Paul Litt, *The Muses, the Masses, and the Massey Commission* (Toronto: University of Toronto Press, 1992) and Kuffert, *A Great Duty*. Tensions between advocates of a public system and the Canadian Association of Broadcasters existed from early on; however, private stations were essential to the life of the CBC. For more on the relationship between private stations and the CBC, see Weir, 224-254.

[98] See the *Annual Report of the Canadian Broadcasting Corporation, 1947-1948* (Ottawa: King's Printer and Controller of Stationary, 1948), 31-33. The Dominion Network was established on January 2, 1944 in an effort to provide more evening options for listeners. See Weir, 233-236.

[99] See Rhodri Windsor Liscombe, "Conditions of Modernity: Si[gh]tings from Vancouver," *Journal of the Society for the Study of Architecture in Canada* 25, no.1 (2000): 3-15 for an excellent discussion of modern architecture's links to postwar social democratic ethos through his comparison of Vancouver's two central branches of its public library system built in 1957 (Semmens Simpson) and 1995 (Moshe Safdie/Downs Archambault). Interestingly, Windsor Liscombe notes that an earlier proposal for a modern library in the business core by Fred Lassarre included an auditorium, office block, and radio station (6).

[100] For more on the history of CKWX, see CKWX-AM in "Radio Station History," *Canadian Communications Foundation*, http://www.broadcasting-history.ca/index3.html?url=http%3A//www.broadcasting-history.ca/listings_and_histories/radio/histories.php%3Fid%3D74%26historyID%3D35 . See also "CKWX Gallery," *Vancouver Radio Museum*, http://www3.telus.net/vanradiomuseum/CKWXGallery.html. After the construction of transmitters on Lulu Island in 1940, CKWX could claim to cover over 100,000 homes in Vancouver, Victoria and surrounding areas. It acquired a shortwave transmitter to beam to more remote areas in the interior and islands of Haida Gwaii.

[101] For more on West Coast Modernism, see Rhodri Windsor Liscombe, *The New Spirit: Modern Architecture in Vancouver, 1938-1963* (Montreal: Canadian Centre for Architecture, 1997).

[102] For more on Broadcasting House see Staffan Ericson, "The Interior of the Ubiquitous: Broadcasting House, London," in *Media Houses: Architecture, Media, and the Production of Centrality*, ed. Staffan Ericson and Kristina Riegert (New York: Peter Lang Publishing, 2010), 19-59; Yusaf, *Broadcasting Buildings*, 19-29; and Elizabeth Darling, *Wells Coates*, Twentieth Century Architects (London: RIBA Publishing, 2012), 99-105.

[103] "CKWX," *Canadian Architect* 2, no. 2 (February 1957): 37.

[104] Liscombe, *The New Spirit*, 102.

[105] Many of the tiles were saved by conservator Andrew Todd and volunteers for the collection at the University of British Columbia in 1989, just before the station was demolished to make way for a new 22-storey condominium development.

[106] CKWX-AM in "Radio Station History," *Canadian Communications Foundation*. CKWX was granted the use of a 50 KW transmitter in 1957, making it the first non-CBC station in the country to transmit with this level of power.

[107] Letter to B.C. Binning from F. H. Elphike, dated Oct. 9, 1956. Cited in note 26 in Abraham J. Rogatnick, "A Passion for the Contemporary," in *B.C. Binning*, edited by Abraham J. Rogatnick, Ian M. Thom and Adele Weder (Vancouver and Toronto: Douglas & McIntyre: 2006), 168.

[108] Culham, 37.

[109] Culham, 37.

[110] H. F. Chevrier, "Canada's Radio Audience - - 1949," *Radio: Staff Magazine* 5, no. 6 (June 1949): 25.

[111] Chevrier, 25.

[112] Parr, *Domestic Goods*.

[113] See Parr, 169.

[114] See Spigel, *Make Room for TV*, for an illuminating discussion of the domestication of television in the USA.

[115] See Kuffert's discussion of *Citizens' Forum*, 83-94. See also Leonard Kuffert, "'Stabbing our spirits awake': Reconstructing Canadian Culture, 1940-1948," in *Cultures of Citizenship in Post-war Canada, 1940-1955*, ed. Nancy Christie and Michael Gauvreau, 27-62 (Montreal and Kingston: McGill-Queen's University Press, 2003).

[116] See, for instance, *Journal, the Royal Architectural Institute of Canada* 24, no. 10 (Oct. 1947) dedicated to school design, as well as the February issue (24, no. 2) of that year, which explores postwar libraries. Superficially, the aesthetic of these public buildings echoes that of the CBC's contemporaneous transmission stations. The school broadcasts are discussed in the CBC Annual Reports.

[117] Parr, *Domestic Goods*.

[118] Marshall McLuhan, "Inside the Five Sense Sensorium," *Canadian Architect* (June 1961): 52.

[119] For more on Canadians' tastes, see Parr, *Domestic Goods*, especially chapters 6, 7, and 8.

[120] "Dine In Style In the Kitchen," *Canadian Homes and Gardens* 26, no. 12 (Dec. 1949): 46, for instance, encouraged readers to design a functional, colour-coordinated eat-in kitchen. Close scrutiny of the images reveals the presence of a miniature radio.

[121] See Susan Douglas' discussion of the shift in radio audiences and listening patterns in American in her chapter "The Kids Take Over: Transistors, DJs and Rock 'n' Roll," in *Listening In*, 219-255.

[122] "10 Ways to Pep Up Your Cottage," *Canadian Homes and Gardens* 26, no. 7 (July 1949): 52.

[123] See Michael Brian Schiffer, *The Portable Radio in American Life* (Tucson and London: University of Arizona Press, 1991), chapter 8.

[124] Schiffer, chapters 9 and 10.

[125] See Schiffer, chapter 12.

[126] See Shapiro, 124-136.

[127] "Save One Corner for Music," *Canadian Homes and Gardens* 26, no. 11(Nov. 1949): 46. Built-in radio was also present in the design schemes represented in *Chatelaine* in the 1950s and 1960s. See Valerie J. Korinek, *Roughing It in the Suburbs: Reading Chatelaine Magazine in the Fifties and Sixties* (Toronto: University of Toronto Press, 2000), 207-210.

[128] Judy Attfield, "Design as a Practice of Modernity: A Case Study of the Coffee Table in the Mid-century Domestic Interior," *Journal of Material Culture* 2, no. 3 (1997): 267-289.

Image Credits

Every effort has been made to acknowledge copyright and/or identify the photographers for all of the images included in this volume.

List of Figures

Fig. 1.1 "Radio!" *Montreal Gazette*, 19 December 1922, 15. Courtesy of Telefonaktiebolaget LM Ericsson.

Fig. 1.2 "The Ultimate in Fine Radio," *Canadian Homes and Gardens*, March 1928, 4. Courtesy of General Electric.

Fig. 1.3 "Splitting Hairs between Toronto and Cincinnati," *Maclean's Magazine*, 1 October 1936, 3. Courtesy of Westinghouse Electric Corporation.

Fig. 1.4 "Great Philcos in 1929," *Maclean's Magazine*, 1 February 1930, 37. Courtesy of Royal Philips.

Fig.1.5 "Northern Electric Mirrophonic Radio," *Montreal Gazette*, 6 December 1937, 15. Courtesy of Telefonaktiebolaget LM Ericsson.

Fig. 1.6 "The Crosley Coloradio,"*Winnipeg Free Press*, 13 April 1951, 4. Courtesy of Royal Philips.

Fig. 2.1 "Tune in with the world of Entertainment!" *MacLean's Magazine*, 14 October 1924, 70. Courtesy of Telefonaktiebolaget LM Ericsson.

Fig. 2.2 "General Electric now presents a 9-Tube Super-Heterodyne," *Winnipeg Free Press*, 15 October 1931, 7. Courtesy of General Electric.

Fig. 2.3 "Radiola Super Heterodyne Receives Stations Thousands of Miles Away Without Aerials or Wires," *MacLean's Magazine*, 15 November 1924, 62. Courtesy of General Electric.

Fig. 2.4 "The Crowning Achievement of a Scientific Century," *Maclean's Magazine*, 15 November 1924, 64. Courtesy of Royal Philips.

Fig. 2.5 "General Electric now presents a 9-Tube Super-Heterodyne" *Winnipeg Free Press*, 15 October 1931, 7. Courtesy of General Electric

Fig. 2.6 "The Children too Can Enjoy Radio," *The Montreal Gazette*, 17 February 1923, 14. Courtesy of Telefonaktiebolaget LM Ericsson.

Fig. 2.7 "Tone: The Product of Radio's Master Minds," *MacLean's Magazine*, 15 September 1926, 3. Courtesy of General Electric.

Fig. 2.8 "Face-To-Face Realism: Big Game Reception that is Truly Alive…" *The Montreal Gazette*, 30 October 1929, 14. Courtesy of Sparton Corporation.

Fig. 2.9 "Face-to-Face Realism: Radio Music That Exalts and Thrills…" *The Montreal Gazette*, 16 October 1929, 16. Courtesy of Sparton Corporation.

Fig. 2.10 "Election Day Monday- August 17th," *The Montreal Gazette*, 14 August 1936, 2. Courtesy of Layton Audio.

Fig. 2.11 "Always a Marvel now a Giant for Power," *The Montreal Gazette*, 25 February 1925, 8. Courtesy of Layton Audio.

Fig. 2.12 "Northern Electric Radio Beauties: These NEW Sets will Amaze You!" *The Montreal Gazette*, 17 November 1932, 2. Courtesy of Telefonaktiebolaget LM Ericsson.

Fig. 2.13 "The Merriest Christmas You've Ever Known," *The Montreal Gazette*, 11 December 1926, 10. Courtesy of Layton Audio.

Fig. 2.14 "Cabinets that Harmonize with Beautiful Interiors," *The Vancouver Sun*, October 14, 1925, 10. Courtesy of Andrea Electronics.

Fig. 2.15 "It's More Fun to Ride with Music," *The Montreal Gazette*, 28 June 1934, 10. Courtesy of General Electric.

Fig. 3.1 "Go Visiting with Viking," *Eaton's Radio Catalogue*, 1937-38, 4-5. Canada Science and Technology Museum trade literature collection L 29175. Used with Permission of Sears Canada Inc.

Fig. 3.2 "Radio Reception in Every Room," *MacLean's Magazine*, 15 May 1929. Courtesy of Telefonaktiebolaget LM Ericsson.

Fig. 3.3 Passengers listen in aboard Maple Leaf radio car, ca. 1929. Canada Science and Technology Museum Corporation / CN Collection: CN000299.

Fig. 3.4 CNR performers in CNRO studio, 1926. Canada Science and Technology Museum Corporation/ CN Collection: CN000301.

Fig. 3.5 "Now—A Radio Set that operates from your Electric Light socket!" *MacLean's Magazine*, 1 November 1925, 25. Courtesy of the E.S. Rogers Family, Toronto, Ontario.

Fig. 3.6 "Complementary Pieces for Additional Groupings," *Canadian Homes and Gardens*, June 1931, 89. Courtesy of Rogers Media Inc., a subsidiary of Rogers Communications Inc.

Fig. 3.7 Whitney Dill, "We Furnish a Four-Roomed Apartment," *Canadian Homes and Gardens*, October-November 1934, 40. Courtesy of Rogers Media Inc., a subsidiary of Rogers Communications Inc.

Fig. 3.8 "Explore the Joys of Home Music," *Canadian Homes and Gardens*, April 1928, 56. Courtesy of Stanley Knight Limited, Beaver Brand Hardwood Flooring.

Fig. 3.9 "Hear this New General Electric Junior Console," *Winnipeg Free Press*, 22 September, 1931, 5. Courtesy of General Electric.

Fig. 3.10 "In this Modern New Plant Marconi has Perfected Radio 3* Different Ways," *The Montreal Gazette*, 8 November 1930, 11. Courtesy of Telefonaktiebolaget LM Ericsson.

Fig. 3.11 W. Gordon Wallace, Cartoon, *Maclean's Magazine*, 1 July 1931, 48. Courtesy of Rogers Media Inc., a subsidiary of Rogers Communications Inc.

Fig. 3.12 "Westinghouse Presents the New Columaire," *Winnipeg Free Press*, 16 May 1931, 16. Courtesy of Westinghouse Electric Corporation.

Fig. 3.13 Kathleen Murphy, "Girth Control," *Maclean's Magazine*, 15 April 1936, 16. Courtesy of Rogers Media Inc., a subsidiary of Rogers Communications Inc.

Fig. 3.14 "Come into the General Electric Home," *Canadian Homes and Gardens*, March 1936, 43. Courtesy of General Electric.

Fig. 3.15 Exterior of CBC station (CBL), Hornby, ON, 1937. D.G. McKinstry (CBC chief architect). From "Canadian Broadcasting Corporation Station, Hornby," *Journal, The Royal Architectural Institute of Canada*, October 1938, 219. Reproduced with permission from the Royal Architectural Institute of Canada.

Fig. 3.16 First floor plan of CBC station (CBL), Hornby, ON, 1937. D.G. McKinstry (CBC chief architect). From "Canadian Broadcasting Corporation Station, Hornby," *Journal, The Royal Architectural Institute of Canada*, October 1938, 219. Reproduced with permission from the Royal Architectural Institute of Canada.

Fig. 3.17 View of interior from visitor's gallery of CBL, showing the control console and map on transmission floor, 1937. Library and Archives Canada, Canadian Broadcasting Corporation Fonds, RG 41, Vol. 542 File Part 1, e010934710. Courtesy of the Canadian Broadcasting Corporation.

Fig. 3.18 CBC broadcast operator at control console. View toward visitor's gallery, CBL transmitter building, 1937. Library and Archives Canada, Canadian Broadcasting Corporation Fonds, RG 41, Vol. 542 File Part 1, e010934717. Courtesy of the Canadian Broadcasting Corporation.

Fig. 3.19 "Normandie," *Montreal Gazette*, 10 November 1936, 7. Courtesy of Telefonaktiebolaget LM Ericsson.

Fig. 3.20 *Philco Radio Atlas of the World*, 1936. Canada Science and Technology Museum Trade Literature Collection L36519.

Fig. 3.21 Transitone, Philco Products Ltd., 1937. Canada Science and Technology Museum Corporation artefact 1977.0511.001. Photo: Peter Coffman

Fig. 3.22 "Just Out! Sensational New Philco Auto Radio!" *The Montreal Gazette*, 26 April 1937, 2. Courtesy of Layton Audio.

Fig. 3.23 Detail from the cover of *Radio: CBC Staff Magazine*, January 1945. Canada Science and Technology Museum Periodical Collection. Courtesy of the Canadian Broadcasting Corporation.

Fig. 3.24 "Seize the Station," *The Montreal Gazette*, 17 June 1942, 7. Courtesy of the E.S. Rogers Family, Toronto, Ontario.

Fig. 3.25 "Out of the Endless Nowhere Radio Gives the 'All's Well'!" *The Montreal Gazette*, 31 March 1942, 7.

Fig. 3.26 "Time out for a cigarette", c. 1945. View of CBC International Service Transmission Station in Sackville, New Brunswick. Library and Archives Canada, Canadian Broadcasting Corporation Fonds, RG 41 vol. 125, file 4 (part 1), e011181977. Courtesy of the Canadian Broadcasting Corporation.

Fig. 3.27 View of main control room of CBC International Service Transmission Station, c. 1945. Library and Archives Canada, Canadian Broadcasting Corporation Fonds, RG 41 vol. 125, file 4 (part 1), e011181978. Courtesy of the Canadian Broadcasting Corporation.

Fig. 3.28 Main Lobby, Radio Canada Building, Montreal, 1951. D. G. McKinstry (CBC chief architect), P.G. Leger (Assistant chief architect). From *Journal, The Royal Architectural Institute of Canada*, March 1952, 65. Reproduced with permission from the Royal Architectural Institute of Canada.

Fig. 3.29 Isometric Drawing, First Floor, Radio Canada Building, Montreal, 1951. D. G. McKinstry (CBC chief architect), P.G. Leger (Assistant chief architect). From *Journal, The Royal Architectural Institute of Canada*, March 1952, 64. Reproduced with permission from the Royal Architectural Institute of Canada.

Fig. 3.30 View into studio, Radio Canada Building, Montreal, 1951. Library and Archives Canada, Canadian Broadcasting Corporation Fonds, RG 41, Vol. 94, 3-16-4 (part 2). Courtesy of the Canadian Broadcasting Corporation.

Fig. 3.31 W. G. Hames, "A Radio Broadcasting Station," *Journal, The Royal Architectural Institute of Canada*, May 1947, 146. Reproduced with permission from the Royal Architectural Institute of Canada.

Fig. 3.32 CKWX Radio Station, Vancouver, BC, 1956. Exterior view. *Journal, Royal Architectural Institute of Canada* (Dec 1958). Reproduced with permission from the Royal Architectural Institute of Canada.

Fig. 3.33 Burrard Street façade, Radio Station CKWX, Vancouver, 1954-56; now demolished. Sharp & Thompson Berwick Pratt; Ron Thom, project architect. Photo: Photographers Ltd. From *Journal, The Royal Architectural Institute of Canada*, December 1958, 445. Reproduced with permission from the Royal Architectural Institute of Canada.

Fig. 3.34 Floor plan and section, CKWX Radio Station. Sharp & Thompson Berwick Pratt; Ron Thom, project architect. From *Canadian Architect*, February 1957. Courtesy of *Canadian Architect*.

Fig. 3.35 "From the Designers of Thorncrest Village," *Canadian Homes and Gardens*, September 1945, 55. Courtesy of Westinghouse Electric Corporation.

Fig. 3.36 "Admiral's Anti-Inflationary Prices Permit You to have a Radio in Every Room!" *Montreal Gazette*, 28 June 1948, 4. Used with permission of Whirlpool Corporation.

Fig. 3.37 "Away you go on a happy Vacation with a new Marconi Portable Radio," *Montreal Gazette*, 17 July 1939, 2. Courtesy of Layton Audio.

Fig. 3.38 "It's light! It's Powerful! It's only $59.95!" *Montreal Gazette*, June 18 1948, 7. Courtesy of Sparton Corporation.

Fig. 3.39 "How to Plan a Good House," *Canadian Homes and Gardens*, Sept. 1950. Courtesy of Rogers Media Inc., a subsidiary of Rogers Communications Inc.

Fig. 3.40 Plate 3.39 "House with 18 sides is full of friendly corners," *Canadian Homes and Gardens*, July. 1956, 12-13. Courtesy of Rogers Media Inc., a subsidiary of Rogers Communications Inc.

List of Plates

Plate 1.1 Marconiphone II, Marconi Wireless Telegraph Co., c. 1923-24. Canada Science and Technology Museum Corporation artefact 1974.0170.001. Photo: Peter Coffman.

Plate 1.2 Front cover, Eaton's *Radio Catalogue*, 1926-1927. Canada Science and Technology Museum Trade Literature Collection L29166. Used with Permission of Sears Canada Inc.

Plate 1.3 Back cover, Eaton's *Radio Catalogue*, 1926-27. Canada Science and Technology Museum Trade Literature Collection L29166. Used with Permission of Sears Canada Inc.

Plate 1.4 "Your General Electric FM Radio Will Bring You Natural Color Music," General Electric Pamphlet, 1944 (no pagination). Canada Science and Technology Museum Trade Literature Collection L05113. Courtesy of General Electric.

Plate 2.1 Front cover of *Eaton's Radio Catalogue*, 1927-28. Canada Science and Technology Museum Trade Literature Collection. Used with Permission of Sears Canada Inc.

Plate 2.2 Front cover of *Eaton's Radio Catalogue*, 1926-27. Canada Science and Technology Museum Trade Literature Collection. Used with Permission of Sears Canada Inc.

Plate 2.3 Front cover of *Eaton's Radio Catalogue*, 1930-31. Canada Science and Technology Museum Trade Literature Collection. Used with Permission of Sears Canada Inc.

Plate 2.4 "Craftsmanship: Only Master Craftsmanship Can Infuse Wood and Metal with this Living Beauty" *Chatelaine*, October 1937, 20. Courtesy of Westinghouse Electric Corporation.

Plate 2.5 "Carry Your Music Wherever You Go," *Chatelaine*, June 1948, 5. Courtesy of General Electric.

Plate 2.6 "Take your "Holiday" anywhere…anytime!" *Maclean's Magazine*, 26 May 1956, 42. Courtesy of Westinghouse Electric Corporation.

Plate 2.7 "Northern's Natural Tone," *Chatelaine*, December 1946, 27. Courtesy of Telefonaktiebolaget LM Ericsson.

Plate 2.8 "Lovely to LOOK at …delightful to HEAR!". *Canadian Homes and Gardens*, October 1953, 89. Courtesy of Telefonaktiebolaget LM Ericsson.

Plate 2.9 "Line of Distinction" advertisement. *Canadian Homes and Gardens*, December 1955, front cover liner. Courtesy of Royal Philips.

Plate 3.1 Radiola III, Canadian Westinghouse Co. Ltd., 1924. Canada Science and Technology Museum Corporation artefact 2001.0219.001. Photo: Peter Coffman.

Plate 3.2 Front cover, Eaton's *Radio Catalogue*, 1924-25. Canada Science and Technology Museum Trade Literature Collection L29163. Used with Permission of Sears Canada Inc.

Plate 3.3 Front cover, Eaton's *Radio Catalogue*, 1925-26. Canada Science and Technology Museum Trade Literature Collection L29165. Used with Permission of Sears Canada Inc.

Plate 3.4 Back cover, Eaton's *Radio Catalogue*, 1925. Canada Science and Technology Museum Trade Literature Collection L29165. Used with Permission of Sears Canada Inc.

Plate 3.5 Rogers "Batteryless" model 110, Standard Radio Mfg. Corp. Ltd., c. 1925-1926. Canada Science and Technology Museum Corporation artefact 1969.0710.001. Photo: Peter Coffman.

Plate 3.6 Model 25-A, Stromberg-Carlson, 1932. Canada Science and Technology Museum Corporation artefact 1969.0688.001. Photo: Peter Coffman.

Plate 3.7 "Baby Grand" Model 20, Philco Products Ltd., 1930. Carleton University Collection. Photo: Peter Coffman.

Plate 3.8 "Columaire," Westinghouse, 1931. Private collection. Photo: Peter Coffman.

Plate 3.9 "Colour Magic that Transforms Your Home," *Maclean's Magazine*, 15 April, 1936, 37. Courtesy of Tarkett Inc.

Plate 3.10 Globetrotter model C-6-1, RCA Victor Co. Ltd, c. 1935. Canada Science and Technology Museum Corporation artefact 1971.0159.001. Photo: Peter Coffman.

Plate 3.11 Front cover, *Eaton's Radio Catalogue*, 1935-36. Canada Science and Technology Museum Trade Literature Collection L40866. Used with Permission of Sears Canada Inc.

Plate 3.12 General Electric Radiotrons Map, 1935. Canada Science and Technology Museum Corporation artefact 1994.0487.001.b. Courtesy of General Electric.

Plate 3.13 RCA Victor Model A-1, 1939-40. Carleton University Collection. Photo: Peter Coffman.

Plate 3.14 RCA Victor A-20, 1940. Carleton University Collection. Photo: Peter Coffman.

Plate 3.15 RCA Victor Radio map, 1940. Canada Science and Technology Museum Corporation artefact 1993.0084.001.a.

Plate 3.16 Northern Electric "Baby Champ" model 5110, 1946-47. Carleton University Collection. Photo: Peter Coffman.

Plate 3.17 "The New Colourful Baby Champ in Six Top Tones," *Chatelaine*, November 1947, 61. Courtesy of Telefonaktiebolaget LM Ericsson.

Plate 3.18 "We asked 240,000 families their preferences in cabinet design," General Electric Pamphlet, 1944 (no pagination). Canada Science and Technology Museum Trade Literature Collection L05113. Courtesy of General Electric.

Plate 3.19 Model A2A, Addison Industries Ltd., c.1946. Canada Science and Technology Museum Corporation artefact 1993.0081.001. Photo: Peter Coffman.

Plate 3.20 "It's Gay—Colourful—Long-Lasting," *Canadian Homes and Gardens*, Sept 1949, 77. Courtesy of Armstrong Flooring Inc., to the extent that Armstrong Flooring Inc. or its subsidiaries own copyright in the above-mentioned.

Plate 3.21 Crosley Model E-15, Crosley Radio & Television Ltd., c. 1953. Canada Science and Technology Museum Corporation artefact 1983.0170.001. Photo: Peter Coffman.

Plate 3.22 RCA Victor model P31, c. 1931-1932. Canada Science and Technology Museum Corporation artefact 1977.0586.001. Photo: Peter Coffman.

Plate 3.23 Hiker, Sparton of Canada Ltd., c. 1947-48. Canada Science and Technology Museum Corporation artefact 1983.0286.001. Photo: Peter Coffman.

Plate 3.24 Continental 5R40, Hallicrafters Canada Ltd., 1953. Canada Science and Technology Museum Corporation artefact 1971.0393.001. Photo: Peter Coffman.

Plate 3.25 RCA Victor Model P-338, RCA Victor Canada, 1959. Canada Science and Technology Museum Corporation artefact 1980.0561.001. Photo: Peter Coffman.

Plate 3.26 Cover, *Canadian Homes and Gardens*, Jan. 1955. Courtesy of Rogers Media Inc., a subsidiary of Rogers Communications Inc.

Plate 3.27 "Electrohome Radio and Television Styled to Compliment your Home," Electrohome Pamphlet, c. 1947. Canada Science and Technology Museum Trade Literature Collection L29159. Courtesy of Electrohome/CWD.

Acknowledgements

This book is the result of over four years of research and work by a number of people. It would not have been possible without the support of the Social Sciences and Research Council of Canada Insight Grant program. We heartily thank our research assistants at York University (Madiha Ahmad, Abiola Akande, Rubina Andaleb, Ebunoluwa Aribido, Laila Balouch, Kadija Bangoura, Kaylin Bean, Matthew Bellefleur, Ruby Boamah, Caterina Borracci, Jean-Michel Busque, Victoria Cake, Alicia Campney, Nicole Ciza-Mugisha, Jazz Cook, Khadija Dairywala, Katherine D'Arolfi, Jemuel Datiles, Lotoya Davids, Andrey Demin, Maximilian Deneau, Michele Deslauriers, Princess DeThomas, Cameron Dick, Julia Dilecce, Nikita D'Souza, Yasmeenah Elzein, Serg Fazylau, Natashia Fearon, Bradley Ferns, Samuel Forrest, Eriene Francis, Ruth Gebremedhin, Laurie Ha, Cindy Ha, Ayan Haibeh, Kristopher Harvey, Jordan Hobbs, Jason Huynh, Habone Imane, David Johnson, Johnesha Johnson, Tamara Khandaker, Zain Khilji, Jinepher Koduah, Jesse Lash, Cindy Le, Sarah Lee, Anna Lendak, Maleeha Maleeha, Christian Maritius, Megan Marshall, Roman Melnyk, Maria Menegakis, Olivia Messina, Jorge Ignacia Miranda Perez, Meagan Miron, Aidan Moir, Alexander Moldovan, Lindsay Moore, Georgette Morris, Jacob Morrow, Kelly Moschopoulos, Jessica Murphy, Ramona Murphy, Chris Naccarato, Leiya Ngo, Bruce Nguyen, Stephanie Nitsos, Emily Norman, Winnie Nwakobi, Kate O'Connor, Yemi Olayiwola, Precious Oppong, Stacy Osas, Paul Owusu-Baah, Alyssa Oziel, Sachin Persaud, Amy Plourde, Kyle Pocalyuko, Modasir Rajabali, Shamar Ralph, Matthew Robichaud, Robyn Rosenblum, Tahmid Rouf, Sandra Roy, Anne-Marie Santerre, Brock Sutherland, Whitney Sweet, Jennifer Terry, Christopher Vanden berg, Aaron Walkins, Nathaniel Weiner, Jessica Whitehead, Doug Wintemute, Sakani Yogeswaran, Zeina Yusuf, Jeffrey Zavala, Melanie Zhang) and Carleton University (Tristan Crawford, Mary Daniel, Gabrielle Doiron, Taylor Gascoigne, Hilary Grant, Laura Hutchingame, Kayla Miller, Paige Pinto, and Marissa Villeneuve).

Special thanks goes to Michelangelo Sabatino and Christine Macy for their interest in including a book about radio culture in the *Canadian Modern* series at Dalhousie Architectural Press. Publications Managers Katie Arthur and Susanne Marshall were essential to the timely completion of the text (and we wish Susanne the best on her maternity leave!).

We would like to acknowledge the support of our home institutions of Carleton University and York University. The encouragement and expertise of our colleagues in the mounting of the exhibitions, especially Alan Steele (Discovery Centre, Carleton) and Sandra Dyck (Carleton University Art Gallery) were indispensible. A huge debt of gratitude is owed to colleagues at the Canada Science and Technology Museum (especially Bryan Dewalt, Tom Everrett, Gordon Perrault, and Catherine Campbell) and to fellow History and Theory of Architecture professor, Peter Coffman, for his wonderful photographs. Thanks to Nancy Duff in the Audio-Visual Resource Centre at Carleton for her assistance with reproductions. Our gratitude and thanks also goes out to Mark Epp, Alison Little and Danielle Manning of the Archives of Ontario as well as Michael Moir of the Clara Thomas Archives and Special Collections of York University for their ongoing support. Thanks also goes to the staff at Library and Archives Canada, Archives of Ontario, Toronto Reference Library, York University Library, Carleton University Library, Vancouver Public Library. We would like, as well, to acknowledge the kind support of all copyright holders.

Finally, we would like to thank our families—Rebecca, Audrey, and Ella; Caeleigh Boara and Shaughna Boara—for their inspiration, support and love. During this project, we both lost our mothers, Ruth Ann Windover and Katherine Ellen Coady MacLennan. This book is dedicated to them.

Notes on Contributors

Michael Windover is Assistant Professor of Art History in the School for Studies in Art and Culture at Carleton University, where he is currently Supervisor of the History and Theory of Architecture Program. He is also Adjunct Curator of Design at the Canada Science and Technology Museum. His research interests include twentieth-century architecture and design culture and the designed environments of mass media. He is author of *Art Deco: A Mode of Mobility* (Québec: Presses de l'Université du Québec, 2012).

Anne MacLennan is an Associate Professor in the Department of Communication Studies at York University and former Graduate Program Director of the Joint Program in Communication & Culture at York and Ryerson Universities. She is the editor of *Journal of Radio & Audio Media*. She has published in the *Journal of Radio & Audio Media, Women's Studies: An Interdisciplinary Journal, The Radio Journal, Relations Industrielles/ Industrial Relations, Urban History Review* and in a variety of collections. Her research interests include community radio, media history, broadcasting (especially radio), popular culture, Canadian history and Canadian studies, women, social welfare, poverty and cultural representations in the media.